Library of Congress Cataloging-in-Publication Data
ISBN 978-0-8311-35034

Industrial Press, Inc.
1 Chestnut Street, South Norwalk, CT 06854
Tel: 203-956-5593, Toll-Free: 888-528-7852
Email: info@industrialpress.com
Web: books.industrialpress.com
ebooks.industrialpress.com

Sponsoring Editor: John Carleo
Interior Text and Cover Design: Janet Romano
Developmental Editor: Robert Weinstein

Benchmarking Best Practices in Maintenance, Reliability and Asset Management:

Updated for ISO 55000

Third Edition

Terry Wireman, CLS, CMRP

WWW.TerryWireman.Com
TLWireman@Mindspring.Com

Industrial Press

Table of Contents

PREFACE

Benchmarking Best Practices in Maintenance Management

Maintenance... many times the word is viewed with a negative attitude. In almost all organizations, the maintenance function is viewed as:

- a necessary evil
- a cost
- insurance
- a disaster repairing function
- prima donnas

As a result, little time or effort is spent on trying to control maintenance activities and costs. Because maintenance has received little budgetary attention in the past, other than a 10% increase or a 10% decrease, most organizations are presently trying to increase profitability. Maintenance expenditures make up a percentage of the cost to produce or the cost of occupancy; therefore, attention is being turned to financial accountability for maintenance expenditures. As organizations audit their maintenance expenses, they find a sizable amount of money spent with little management control.

Proper management controls must be applied to maintenance if costs are to be curbed. But to successfully control maintenance, proper management policies and practices must be instituted. Again, many organizations have tried to use standard production or facilities oriented methods to control maintenance. This has not and will not be successful. Maintenance is a unique business process. It requires an approach that is different from other business processes if it is to be successfully managed. It is the purpose of this text book to present insight into what is required to manage maintenance properly. The text does not provide a total answer to every maintenance management problem. However, it will provide a framework with options, allowing maintenance decision makers to select the most successful way for them to manage their business.

The Business of Asset Management

The first step is to understand where maintenance fits in the overall business process. An international effort to develop an ISO standard for asset management was just completed in January of 2014. Because it has just been completed (as of the writing of this text), it will take some time for many companies to decide to pursue certification under this standard or to wait until it is mandated.

If an ISO standard is to receive wide acceptance by the international asset management community, it will have to have a positive impact on a company's overall business. If not, then it will be a standard that some can point to, but senior executives will not worry about compliance. The only way that the ISO-9000 (Quality) standards gained worldwide acceptance is when compliance with it became a business requirement. Once senior executives realized the importance it had to their continued profitability, they mandated their companies become compliant.

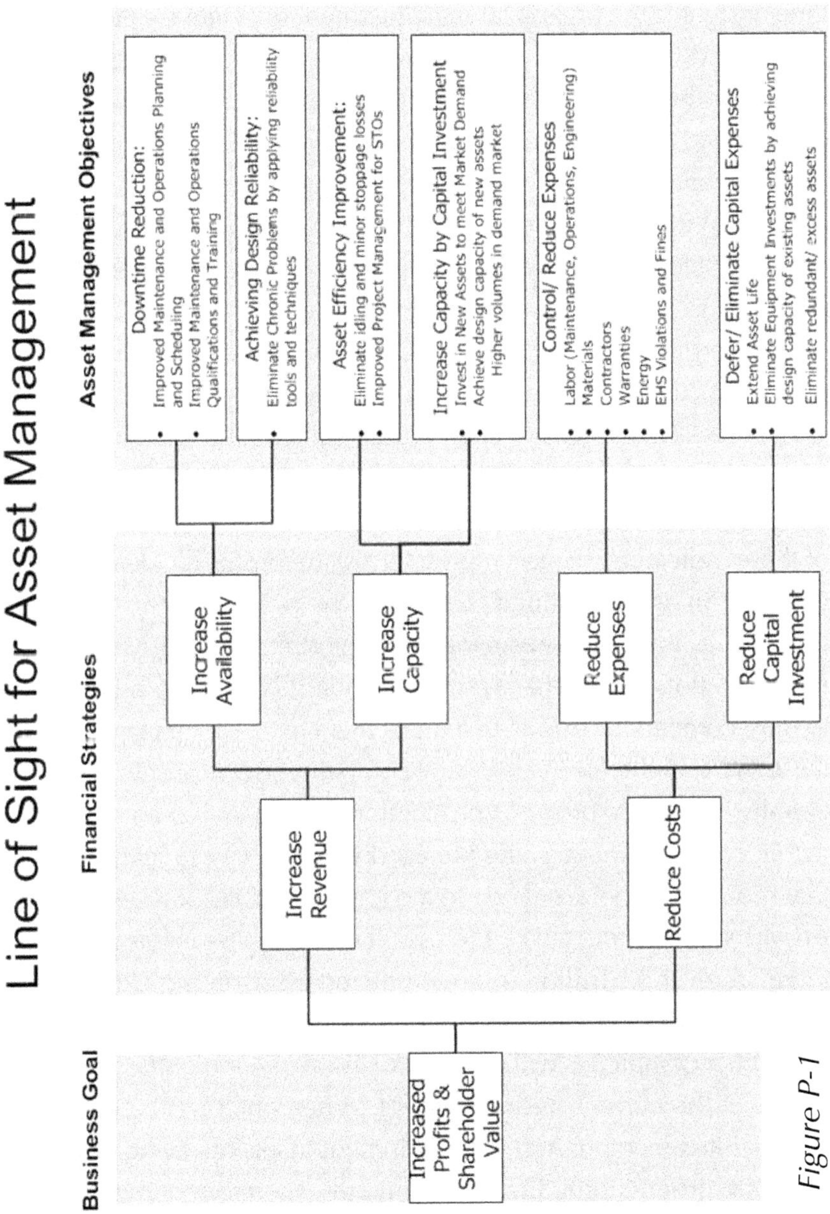
Line of Sight for Asset Management
Business Goal
Financial Strategies
Asset Management Objectives
Increased Profits & Shareholder Value
Increase Revenue
Reduce Costs
Increase Availability
Increase Capacity
Reduce Expenses
Reduce Capital Investment
Downtime Reduction:
Improved Maintenance and Operations Planning and Scheduling
Improved Maintenance and Operations Qualifications and Training
Achieving Design Reliability:
Eliminate Chronic Problems by applying reliability tools and techniques
Asset Efficiency Improvement:
Eliminate idling and minor stoppage losses
Improved Project Management for STOs
Increase Capacity by Capital Investment
Invest in New Assets to meet Market Demand
Achieve design capacity of new assets
Higher volumes in demand market
Control/ Reduce Expenses
Labor (Maintenance, Operations, Engineering)
Materials
Contractors
Warranties
Energy
EHS Violations and Fines
Defer/ Eliminate Capital Expenses
Extend Asset Life
Eliminate Equipment Investments by achieving design capacity of existing assets
Eliminate redundant/ excess assets

Figure P-1

This is why the "line of sight" for any business is so important. Line of sight is the organizational alignment to achieve a goal. It is focusing and then connecting a company's business processes to accomplish the company's overall business objectives. This could be compliance to governmental regulations, which is important to many organizations. However, for most companies, line of sight is going to be profitability. For these organizations, it will be the answer to the question "How will achieving an asset management standard help my company to become more profitable?"

For profit oriented companies, line of sight can be illustrated by a diagram such as Figure P-1.

Increasing profits and shareholder value is the reason that companies are being measured by indicators such as ROIC (Return on Invested Capital) or ROA — (Return on Assets). In simplest terms, both of these indicators (in the simplest terms) are profits divided by the investment in assets required to provide the profit, which most shareholders are interested in seeing. The investment in the assets is why an international asset management standard is going to be important to most companies. True asset management means to manage the asset through the entire life cycle. Many companies overinvest in assets and literally have spare process / production lines in their plants / facilities just in case the primary line breaks down. We can quickly see what that business model would do to a company's ROIC or ROA indicator. An MIT study conducted by Professor Erik Brynjolfsson has shown that there are over $2 Trillion in underutilized assets in the United States alone — what would that number be worldwide?

Other companies will design and install equipment/assets that do not meet the current business needs. When entering the operations and maintenance phase of the equipment/assets life cycle, it will not meet the production/utilization demands; therefore companies will push it beyond its design capacity. This results in the equipment/asset being unreliable, breaking down more frequently, and wearing out prematurely. This also has a severe impact on the ROIC or ROA indicators. Capital dollars are then re-invested in rebuilding the equipment to bring it back to an acceptable baseline for some measure of the rest of its (now shortened) life cycle.

This should highlight the fact that when a company uses the "line of sight" term, they have to be focusing on achieving business objectives, which for the majority of organizations, is going to be increasing shareholder value by increasing profitability. When organizations examine the benefits of implementing an asset management standard, almost all of the existing data shows that the majority of the financial benefits are derived from improving maintenance and reliability practices. If a company focuses on achieving an asset management standard, it should steer that standard in a direction that improves their businesses; otherwise there is no business reason to adopt it.

Asset Life Cycle Management and Asset Management Standards

Why is the terminology Asset Life Cycle important to the development of an asset management standard? If companies are going to adopt an asset management standard that has any real meaning, it has to be built on the documented science of asset management. There are literally hundreds of facility management, plant management, engineering design, and other related textbooks that discuss this topic. There are scores of engineering courses taught world-wide on this topic. If an asset management standard does not contain accepted engineering principles, the acceptance of the standard will be diminished by the business communities that will benefit the most from accepting the standard.

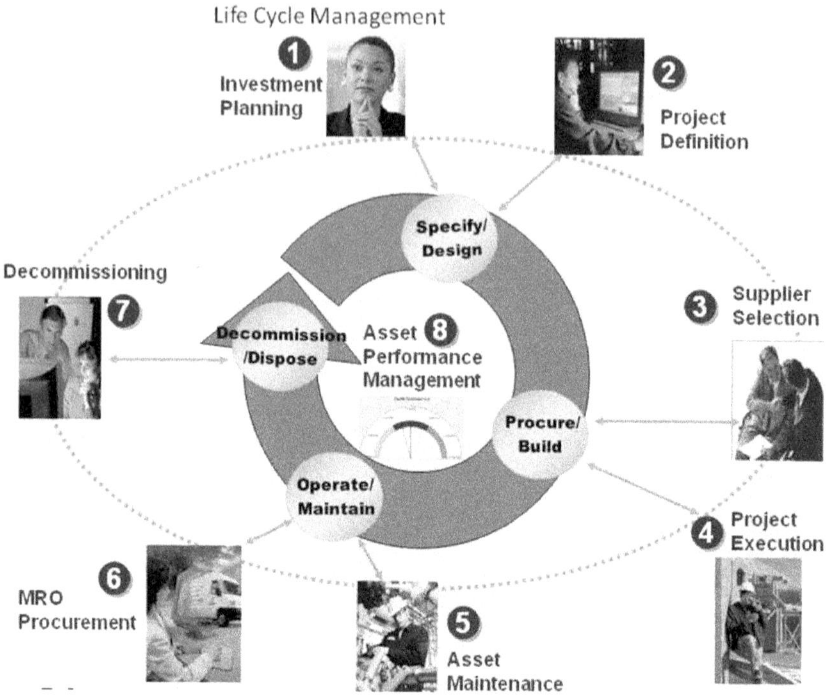

Figure P-2 highlights the main phases of an asset life cycle.

The following material will review each major phase (from Figure P-2) of an asset's life cycle.

1. Investment Planning (Needs and Feasibility Assessments for Assets)

This phase of an asset's life cycle begins with the discovery that there is a new product or service that can be produced and sold, there is a greater demand for an existing product or service has a greater demand, or another facility location is required. The demand for new assets may also relate to meeting increased regulatory requirements for existing assets. This may involve:

a. Strategic Planning

i. The company direction is to diversify, expanding into new markets

ii. The company direction is to expand their share of an existing market

b. Customer Needs

i. The customer demands modifications or enhancements to existing products or services that requires new assets

c. Regulatory Requirements

i. There may be new regulatory requirements that require extensive modifications (new assets) to existing buildings, facilities, process equipment, etc.

2. Project Definition (Design of Assets)

In this phase of an asset's life cycle, the scope of the asset is defined. For the asset to meet the demand (identified in phase 1), it will need to meet certain requirements. There are certain reliability, maintainability, projected life, and total cost of ownership (TCO) requirements that all assets will need to meet to support the business requirements identified in phase 1. For example, how long will the company occupy the facility? How many people will occupy the building? What is the volume of air that the HVAC system must move to make the building comfortable? What is the thermal load that the HVAC equipment must be able to handle in the four different seasons? The list can be quite extensive. (Note: There are a considerable number of texts written on this topic. One is *Facilities Operations and Engineering Reference* published by R.S. Means, in 1999).

In addition to a facilities example, there are others that need to be considered. For example, what is the business need is met by a new production line? A new process? These assets will also have certain design reliability, maintainability, projected life, and TCO requirements that all assets will need to meet in order for it to support the business requirements identified. What is the production volume that must be achieved to meet the business need identified in Phase 1? Will the assets be required to perform in 7 X 24 operations or will it be a 5 x 24 schedule? The reliability (how long the equipment operates in between maintenance periods) and maintainability (how long it takes to restore the equipment to service) are critical to the decision on the capacity of the asset and the profitability (ROIC or ROA) of the new product/ service.

This leads to the cost-benefit analysis. Will the company specify a facility building that is designed for 500 people, when the business plan requires 100 employees? Conversely, will the company design a building for 500 people, when the business plan requires 1000? Both mistakes will be expensive. When considering production assets, if the assets need to produce 1000 bottles of beer per hour, will the company design a line that produces 10,000 bottles of beer an hour? Or will it design a line that is only capable of producing 500 bottles of beer per hour? Any mistakes in designing assets, where the design is not based on the company's long-range strategic plan will result in extreme financial penalties for the company.

It must be kept in mind that the asset, at this phase of its life cycle, is still only a document, a drawing, or a blueprint. There have been no major costs (other than studies) done to this point. In fact, there are dozens of texts written on life cycle costing that show up to 90% of the life cycle costs are specified (knowingly and unknowingly) by the asset design engineer. However, the same 90% of the assets life cycle costs are not incurred until the asset is in its operational and maintenance phase of the life cycle. Unfortunately, most companies overlook this fact and fail to achieve the profitability projections in the Phase 1 business study.

There are three calculations that a design engineer will focus on during the design of an asset. They are:

a. **Reliability** — A design specification that determines the period of time an asset will perform its intended function without failure. This is typically measured by the Mean Time Between Failure (MTBF) calculation.

b. **Maintainability** — A design specification that determines the length of time it takes to restore an asset to its functional state once a failure has occurred. This is typically measured by Mean Time To Repair (MTTR) calculation.

c. **Cost-Benefit Analysis** — A design study that shows the profits required from an asset versus the cost the asset will incur throughout its life cycle. It may be measured the ROIC or ROA calculations.

How does the asset move from the drawing board to the plant floor? How do these calculations impact the asset during the rest of its life? How does a company maximize their return on investment in the asset?

3. Construction, Acquisition, or Enhancement of Assets

In this phase of an asset's life cycle, it is actually created, produced, or acquired. The initial construction/acquisition cost is actually incurred at this time. If the asset is constructed internally, all of the design documents, capacity studies, reliability and maintainability specifications, regulatory requirements, etc., are utilized to construct an asset that will provide the company with the maximum ROA/ROIC for their shareholders and stakeholders. This is the case, whether it is a building, a process, or a production line that is being constructed.

If the asset is to be purchased, all of the same design documents, capacity studies, reliability and maintainability specifications, regulatory requirements, etc., are provided to the vendor who will be constructing or providing the new asset. The company will audit the delivered asset against the specifications to insure the proper asset has been supplied by the vendor.

If existing assets are to be redesigned or modified to meet the business plan, then all of the same specifications that would have been developed for a new asset (whether constructed internally or purchased externally) are used during the modification of the existing asset. At the end of the redesign or modification, the asset should be capable of delivering its design capacity at specified cost.

4. Project Execution of the Assets

In this phase of the asset's life cycle, the asset — whether it is built, purchased, or retrofitted — is installed in the plant or actually built. This is a construction or installation project. There is some divergence, based on the philosophical leaning of the author, professor, or practitioner. Figure P-3 is used by Professor Ben Blanchard in his Engineering Masters program.

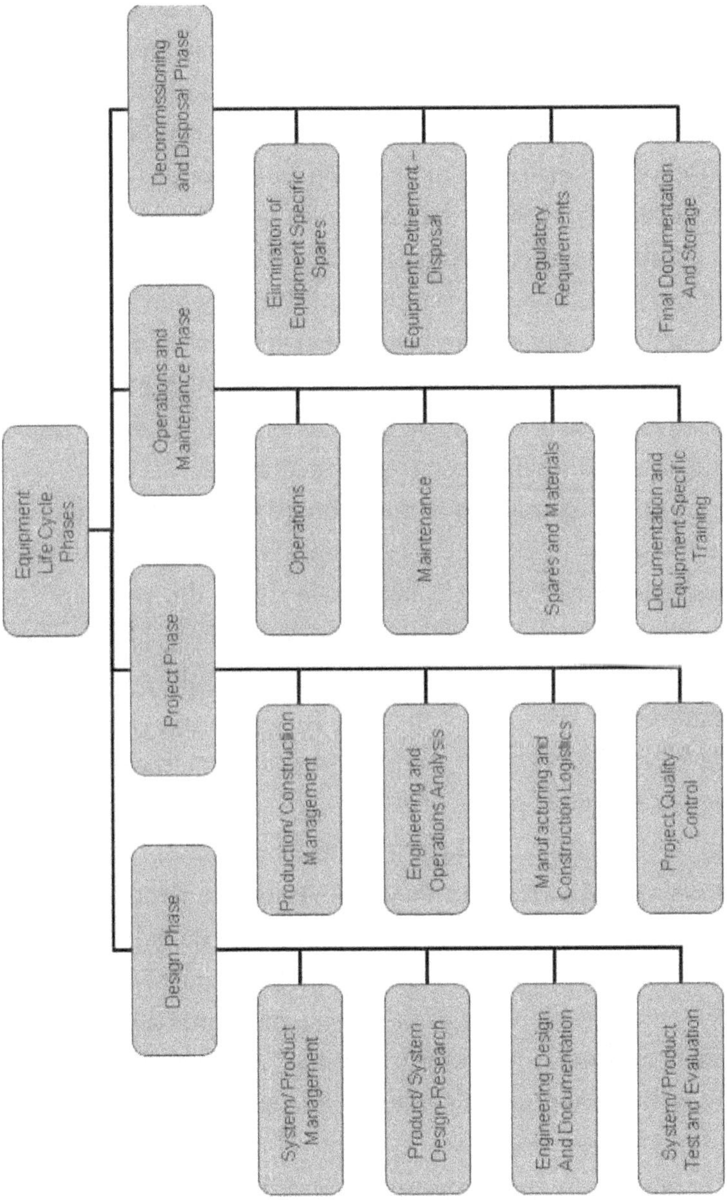

Based on Life-Cycle Cost And Economic Analysis –
Benjamin S. Blanchard
(Professor of Engineering-Emeritus
at Virginia Polytechnic Institute and State University)

Figure P-3

This diagram rolls the construction, acquisition, and installation of the asset into one phase. However one would choose to combine or separate these phases, the project phase involves the installation of the equipment, or construction of the building. This phase is critical because poor installation and construction practices can diminish the design reliability and maintainability of the asset. For example, poor foundations under equipment can make it virtually impossible to achieve its reliability and maintainability design specifications.

During this project phase, commissioning also occurs. In a building setting, the final inspection and walkthrough occur before the asset ownership transfers from the builder to the company purchasing the building. In an equipment setting, all of the asset's capacities are tested to insure they meet all of the design specifications.

Once the commissioning phase is achieved, the asset ownership passes from the supplier (internal or external) to the company. All documents, manuals, drawings, etc., are transferred to the company. In many cases, all of the documentation is provided to the company electronically. This may also include the requirement for the supplier (internal or external) to enter all of the data for the new asset into the company's CMMS/ EAM system. At this time, the asset moves into the maintenance / operations phase of its life cycle.

5. Operations and Maintenance Phase of Asset Life Cycle

This phase is where the asset actually starts providing the business service or production that was envisioned back in the first phase of its life cycle. It is much more than allowing occupants into a building or pushing a button to turn on a production line / process. It is insuring that it delivers the design specifications. There are two aspects to this: Operations and Maintenance.

From an operations perspective, does it deliver the design capacity and capabilities? For example, use the building example: Can it hold all of the occupants it was designed to accommodate? Is the HVAC adequate to keep everyone comfortable? Do the restrooms provide satisfactory service? Is the lighting sufficient? If any of these systems fall short, the capacity of the building is diminished. From an equipment perspective, does it achieve the design capacity? Does it produce at the rate for which the original design specified? Does the asset meet the "operability" that was specified in the initial design documentation?

From a maintenance perspective, the asset was designed with certain reliability and maintainability parameters. Does the equipment achieve the design Mean Time Between Failures (reliability) and the design Mean Time To Repair (maintainability)? If not, why not? Do the specified maintenance policies and procedures insure that the design specifications can be achieved? If not, why not? It was mentioned previously that up to 90% of the asset's life cycle cost will be incurred during this phase of the asset's life. These costs are commonly divided into labor costs, materials costs, and contractor costs.

Some companies will try to reduce these maintenance expenditures to a level where the asset is not properly maintained. When this occurs the asset will either fail to produce the design reliability or maintainability. In some cases catastrophic failures occur, creating a negative image of the company (and financial penalties which have involved numerous companies lately). Due to insufficient maintenance expenditures, particularly in buildings, they will fail to achieve their full design life. This leads to premature major overhaul/ refurbishing or even decommissioning expenditures. In these cases, the original ROA / ROIC that was promised to shareholders and stakeholders will never be achieved. Eventually share prices in the company's stock will likely suffer.

In this phase of the life cycle, information management is critical. Tracking all of the data in a CMMS/ EAM system insures that the proper level of maintenance activities are being performed on the assets to insure they deliver their design functions at the designed costs. This data is essential in a later phase of the life cycle if any business improvements related to the assets are to be achieved.

6. Rehabilitation or Retrofitting of Assets

Some texts, professors, companies, etc., will treat this section as a separate life cycle phase, whereas others will incorporate it into the maintenance and operational phase. The activities surrounding shutdowns, turnarounds, and outages are focused on restoring deteriorated assets to an acceptable base line to insure that the original design capacities can be achieved. In the cases of buildings, they are refurbished to bring any deterioration to an as new condition assuring the structure is fit to purpose.

Some organizations will move reengineering and retrofits of existing assets into this phase of an asset's life cycle. Others will include it in the maintenance and operations phase. In this case, workers reengineer an older asset and upgrade it to current market standards, ideally so that it meets new building codes or new production specifications.

Assets that need to be replaced move into their final life cycle phase.

7. Decommissioning, Retirement, and Disposal of Assets

In this phase of an asset's life cycle, it is time decommission and dispose of it. In some cases, decommissioning is nothing more than writing it off the books and gradually selling off components of the process or manufacturing line. When a process or manufacturing line is decommissioned, it may be cannibalized for parts that can be used for other assets that are still operating elsewhere in the plant or facility. Ultimately there will be a time that the remaining components are sold for scrap.

Another popular option for disposing of an asset is to sell it to another company. This occurs frequently when a company has decided to divest a certain product or a certain part of its current market. Another company will purchase the asset, may retrofit it and continue production of the product or a similar one.

In most cases, however, a facility (such as a building) will likely not be cannibalized for parts. Instead, it will be sold off as a method of disposal. Vacant structures often quickly become a legal liability. This encourages a company to sell the asset rather quickly.

Depending on the asset, there may also be regulatory considerations when it is decommissioned. If it was part of an environmentally hazardous process, then there will be regulations that must be met. Depending on the country in which the asset is located, the range of remediation activities will vary dramatically. In the United States, there are regulations from OSHA, EPA, and various state and local regulations that must be met.

Eventually, through whatever process a company has in place, the asset is decommissioned, written off the company register, and finally disposed.

8. Performance Management

This aspect of an asset's life cycle is often overlooked. However, if you consider the center of Figure P-2, performance management is critical throughout the life of the asset. Although there are many different aspects of performance management, if a company does not measure the performance of an asset, it cannot be improved in the next generation of asset evolution.

Performance management is a key. For example, each company asset was originally conceived (as part of the asset base to provide a new product or service) and designed to perform a task in the plant or facility. Some performance measures must audit the installed and operating asset against what it was conceived and designed to perform. Whether it was a building that was supposed to house 500 people for 8 hours a day for 5 days a week for 40 years or a pump that was supposed to provide 100 gallons of flow per minute for 24 hours per day, 7 days a week, for 15 years, the question must be asked "did it achieve that level of design performance?"

If it did, then the performance was satisfactory. If it did not, then why not? This is where a typical CMMS / EAM system is used to monitor performance. A typical system will track the mean time to failure (MTBF) and the mean time to repair (MTTR). It tracks the labor and material costs (whether internal, contractors, or a combination) that the asset required to keep it at design performance.

Although not getting into a high-level specificity, there are multiple aspects that need to be measured during the asset life to enable future improvements. After this information has been tracked through the life cycle, it is fed back into the "conceive" and "design" phases of the next generation of assets for the company. Any weaknesses in the prior assets are corrected in the new design or specification for the next generation of assets. Without this performance measurement feedback loop, companies will not be able to improve their competitive positions in their respective marketplaces.

Figure P-4 provides another way to look at questions to be asked during an asset life cycle.

Up to 95% of the asset's life cycle cost is specified in the design and installation (procure and build) phases. The majority of the costs

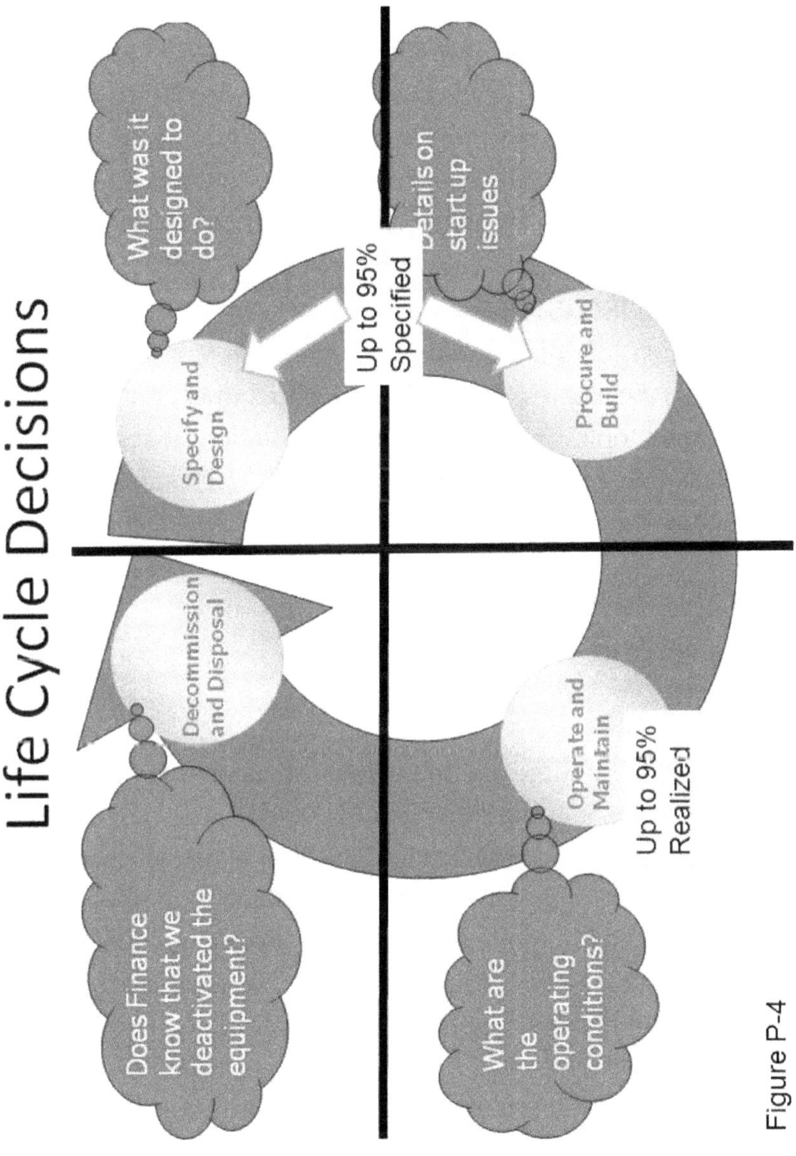
Life Cycle Decisions
What was it designed to do?
Details on start up issues
Up to 95% Specified
Specify and Design
Procure and Build
Decommission and Disposal
Operate and Maintain
Up to 95% Realized
Does Finance know that we deactivated the equipment?
What are the operating conditions?

Figure P-4

are not actually incurred until the operational and maintenance phases. So when an organization benchmarks with another company, there may be significant differences between their maintenance expenditure and yet is it not the fault of the maintenance organization. It could be the design or project engineering function that is driving the costs higher. This fact should have a major influence on the maintenance activities that are included in any benchmarking project.

PAS-55 AND ISO-55000

After considering the role of maintenance and reliability in the life cycle of the assets, it is also important to further consider the asset management topic. There are three key definitions that are important to understanding the PAS-55 and ISO-55000 asset management standards. They are:

- Asset
- Asset Management
- Asset Management System

The IAM (www.theiam.org) defines an asset as "things that have actual and potential value." The ISO-55000 document defines an asset as an "item, thing, or entity that has potential or actual value to an organization." Although both of the standards focus on physical assets, the concept can be applied to all assets, including financial that an organization possesses. The real issue is how do you manage the assets to provide the maximum value that they can to the company or organization?

This is where the term asset management develops meaning. Asset management converts the business plan for the organization into coordinated activities that realize the maximum value from the assets.

Properly managing assets will require the organization to document their business objectives and convert that into asset management objectives. In ISO 55000, this documented plan is referred to as the strategic asset management plan.

The asset management system is a set of interacting elements that will establish the asset management policy, the asset management objectives, and all of the processes necessary to achieve those objectives. The asset management system specifies:

Resources — To enable the asset management objectives to be achieved; financial, man-power, etc.

Competencies — Insures that persons involved in asset management are competent based on their education and training.

Awareness — How persons doing work can impact achieving the asset management objectives.

Communication — The necessary communication to internal and external stakeholders relevant to the assets, asset management, and the asset management system

Information — The information necessary to support the assets, asset management, and the asset management system

Documentation — The documentation necessary for an organization to be in compliance with the ISO-55000 standard.

Each of these processes must be specified for the asset management system to allow the organization to achieve their asset management objectives.

Within the asset management system, the asset life cycle activities are specified. These are the activities that are applied to the assets during their life cycle to maximize the value that they provide the organization.

The relationship among these terms is pictured in Figure P-5, which is a draft form of the graphic that is found in the ISO-55000 document.

Figure P-5 shows that the overall business plan or managing the organization drives the need for asset management. With asset man-

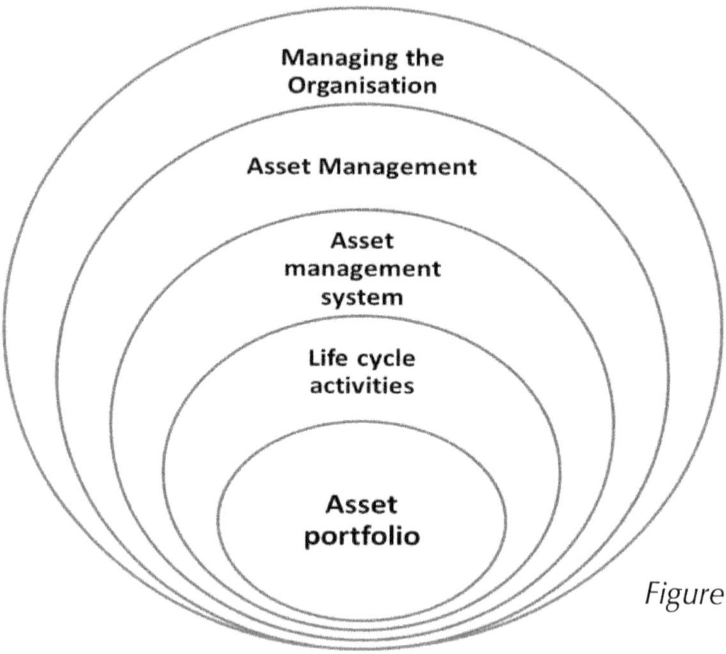

Figure P-5

agement goals and objectives clearly understood, an asset management system is developed to insure the organization meets the goals of asset management. Contained within the asset management system are the activities that are performed directly to the assets during the life cycle stages. And at the center is the asset portfolio for all of the assets that are within the scope of the asset management system.

Although asset management is much more than just maintenance and reliability, many of the activities are the focus of maintenance and reliability. As highlighted previously, maintenance and reliability play a large role in the life cycle of the asset. In Figure P-5, the life cycle activities circle is where maintenance and reliability will have their greatest impact. The relationship between maintenance and reliability and all of the other asset management related business function in an organization must be clearly understood if any benchmarking effort is to be successful.

Conclusion

Why is this information important to any organization considering benchmarking their maintenance and reliability functions? It is important because so many organizations want to benchmark their maintenance and reliability costs. If the costs are specified in the design and project phases of an asset's life cycle, then these activities should also be noted during a benchmarking study as a hidden factor that contributes to the maintenance costs. This is especially true if the maintenance costs tend to be higher than any of the benchmarking partners in the project.

Understanding the impact that life cycle decisions have on the maintenance function should always be a factor in any maintenance benchmarking project.

Foreword

Status of Maintenance in the United States

In the United States, it was estimated that in 1979 there were over $200 billion spent on maintenance. This is a sizable figure in anyone's estimation. However, more disturbing than the amount was the fact that approximately one-third of the total was spent unnecessarily (wasted). As we continue to the present, there have been no significant changes in maintenance policy, indicating the waste trend is probably still about one-third. The largest change in the maintenance costs is the amount. Since 1979 maintenance costs have risen between 10% and 15% per year. If this amount is calculated, it is very likely that the maintenance expenditures in the United States are now over $1 trillion per year. If the waste ratio is holding steady, and there are no indications that the ratio is changing, it is likely that companies are wasting many times over what they were spending on maintenance 35 years ago. These costs are illustrated in Figure F-1. Where do these wastes occur in maintenance? How can they be controlled? These questions can best be answered by looking at some statistics.

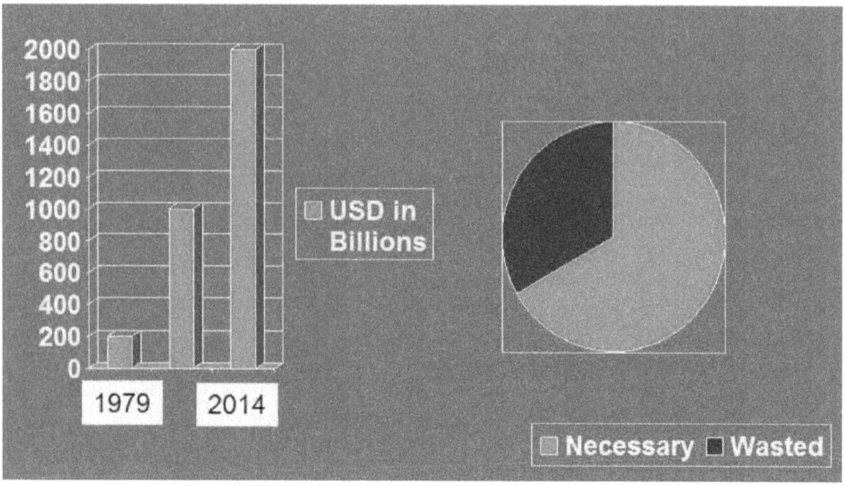

Figure F-1, Maintenance Cost Escalation

1. Less than 4 hours per day (out of a possible 8) is spent by maintenance craftsmen performing hands-on work activities (See Figure F-2). This fact is even more alarming when it is realized that the majority of maintenance organizations are as low as 2 hours of hands-on work (or wrench time). It is not that these employees are lazy or shirking job responsibilities. It is the fact that they are not provided the necessary resources by management to perform their assigned job tasks. Providing these resources becomes important to increasing maintenance productivity and producing a substantial maintenance labor savings. If one would view maintenance salaries as a resource they are paying (for example) $20.00 per hour and only utilizing at a 50% level, there is a tremendous cost waste. Methods that may be used to increase labor productivity will be explained in Chapter 6.

2. Only about two-thirds of all maintenance organizations employ a job planner to plan maintenance activities (see Figure F-3). How-

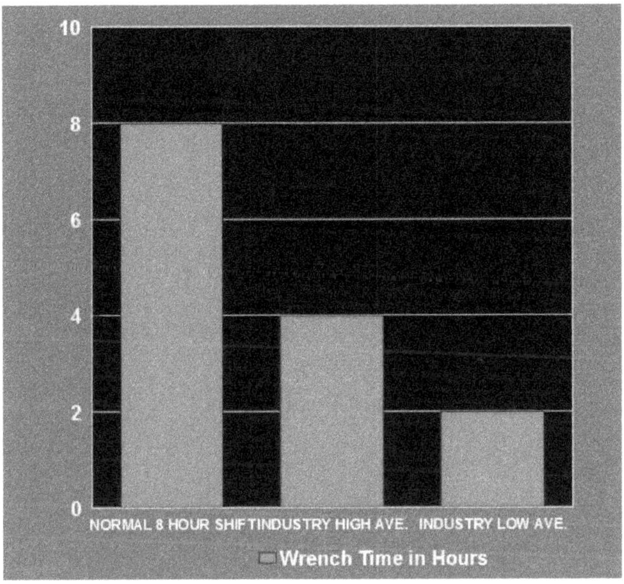

Figure F-2, Actual Hands-On Work

ever, most organizations use unclear job descriptions or fail to maintain a proper craft technician-to-planner ratio. Most experts agree this is one of the largest potential areas for cost savings in the maintenance arena because maintenance planners insure that maintenance work is performed effectively and efficiently. It is estimated that planned versus unplanned work may have a cost ratio as high as 1:5. Performing a $100 planned job could save as much as $400 over performing the same job in an unplanned mode. We will explore maintenance planners and their qualifications and assignments in Chapter 6.

3. The majority of all maintenance organizations either are dissatisfied with or do not have work order systems (see Figure F-4). This is one of the most critical indicators of the status of a maintenance or-

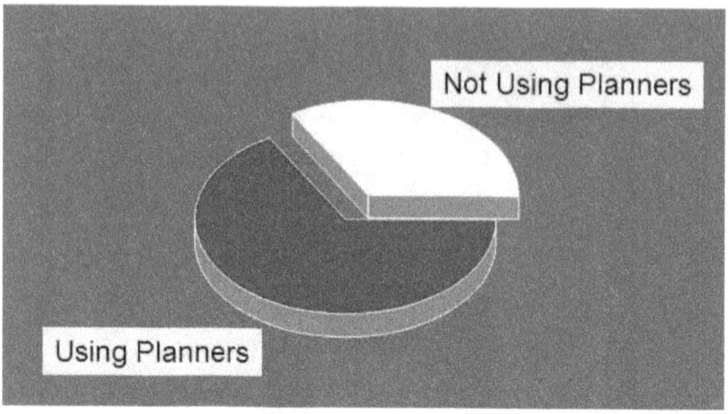

Figure F-3, Maintenance Planners Perspective

ganization. If a maintenance organization does not have a work order system in place that works correctly, it is impossible to measure or control maintenance activities. The importance of a maintenance work order system and how to set up and use the system will be discussed in Chapter 5.

4. Of companies that have work order systems (1/3 of all companies), only about 1/3 track the work orders in a craft backlog format

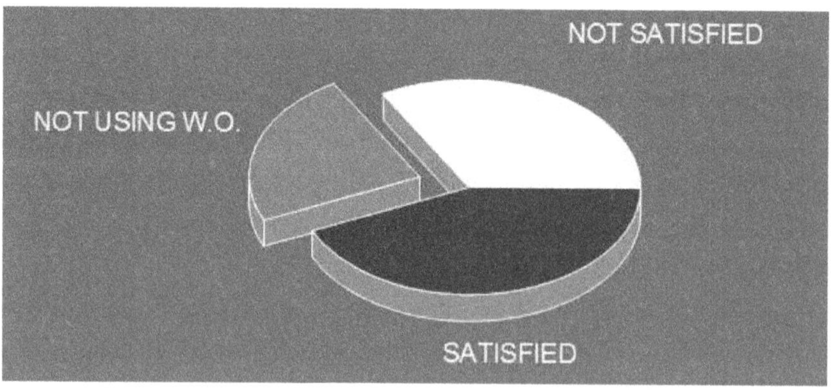

Figure F-4, Work Order System Analysis

— actually 10% of total organizations (see Figure F-5 and F-6). The craft backlog allows managers to make logical staffing decisions based on how much work is projected for each craft. It will be a departure from the "I think I have enough craft technicians" or "let's work overtime to get caught up" modes that most organizations find themselves in today. Being able to justify employment levels to senior management is a necessary function of good maintenance management. We will discuss backlogs and their significance in Chapter 6.

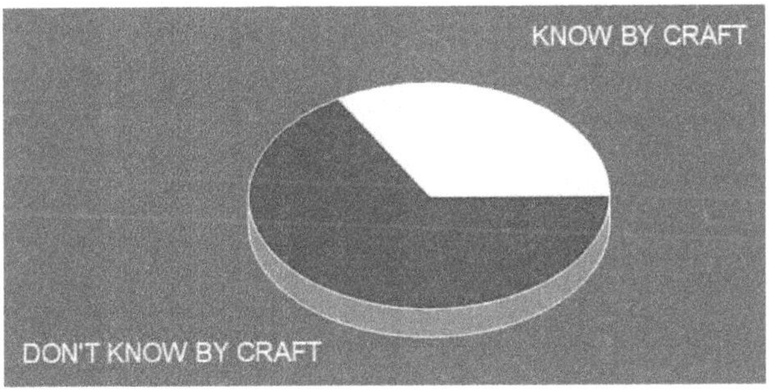

Figure F-5, Craft Backlogs

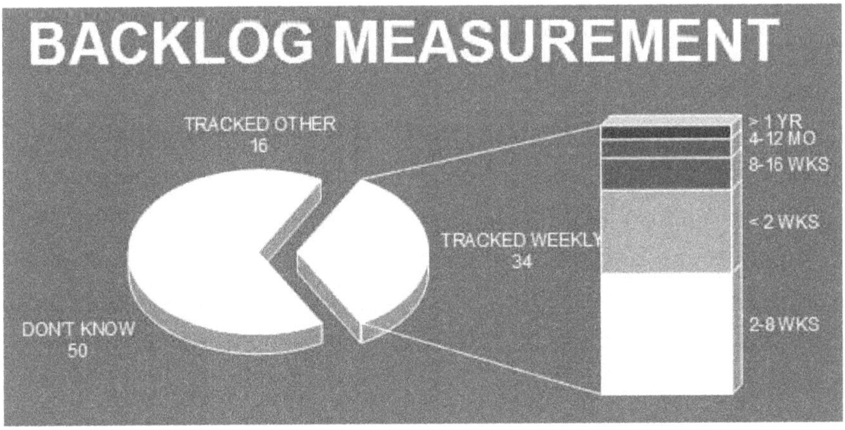

Figure F-6, Backlog Measurement

5. Of companies that have work order systems, only one-third compared their estimates of the work order labor and materials to the actual results (see Figure F-7). This means again only about 10% of all organizations carry out some form of performance monitoring. Successful maintenance management requires performance monitoring. Proper methods of performance monitoring including maintenance analysis will be covered in Chapter 9.

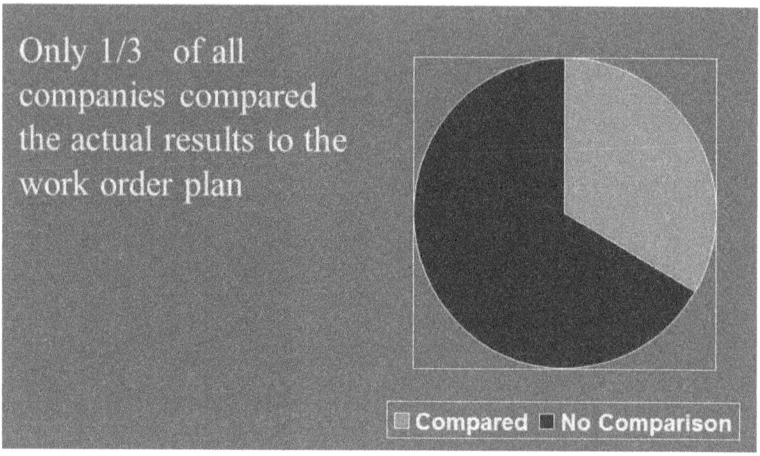

Figure F-7, Work Order Completion Analysis

6. Of the companies with work order systems that allow for feed back, only one third, again about 10% of all companies, perform any failure analysis on their breakdowns (see Figure F-8). Most of the other companies are just parts changers. For an operation to be cost effective, good practice in failure analysis (commonly called Root Cause Analysis or RCA) must be followed. We will discuss this topic in Chapter 9.

7. In the United States, overtime, another key indicator, averages about 14.1% of the total time worked by maintenance organizations

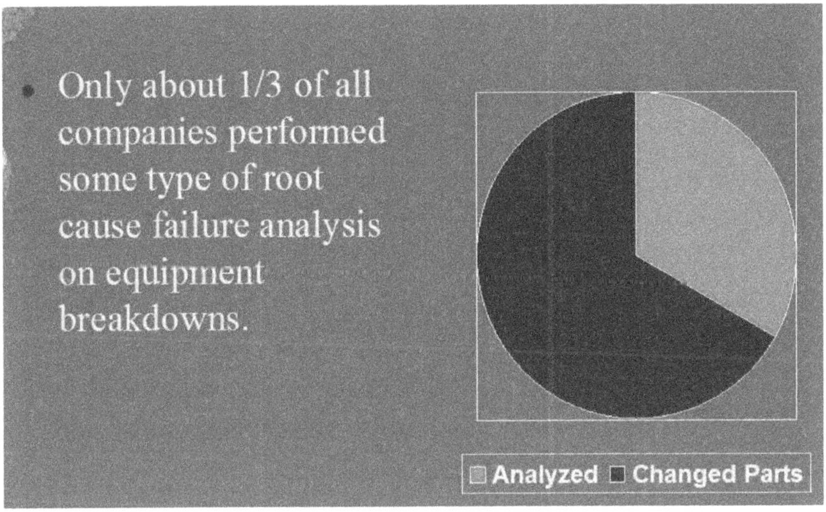

Figure F-8, RCA

(see Figure F-9). This level is almost three times what it should be. Because maintenance is working so much overtime, it again indicates the reactive situation that is standard in the maintenance process. Reducing overtime is essential if a maintenance organization is to be truly cost effective. Proven methods for reducing overtime will be explored in Chapters 5 and 6.

8. Preventive maintenance, another major part of any successful maintenance program is presently satisfying the needs of about 22% of the company's surveyed (see Figure F-10). This again shows major problems for the maintenance organizations. Without successful preventive maintenance programs, maintenance can only react to given situations. Preventive maintenance allows the organization to be proactive. This means more planning and reduced maintenance costs. Over three-fourths of the organizations show needs for major improvements. Methods of implementing and improving preventive

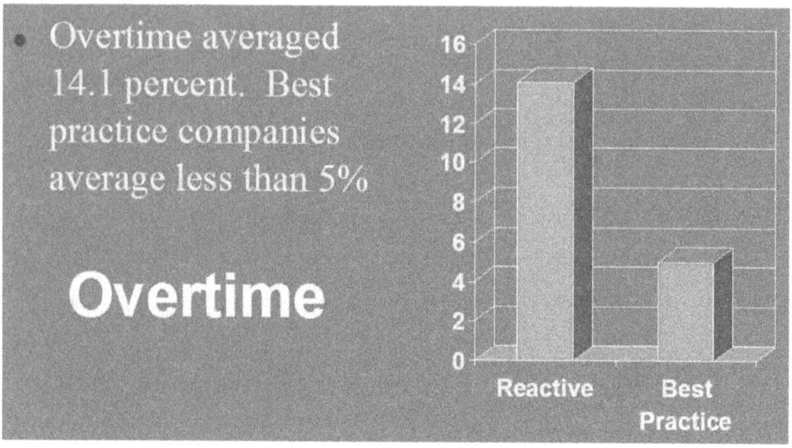

Figure F-9, Overtime

maintenance programs will be discussed in chapter 7.

9. Related to preventive maintenance, almost three-fourths of the organizations have some form of lube routes and procedures (see Figure F-11). Although this fact seems to be positive on the surface, it is not. Many of the organizations feel that preventive maintenance is

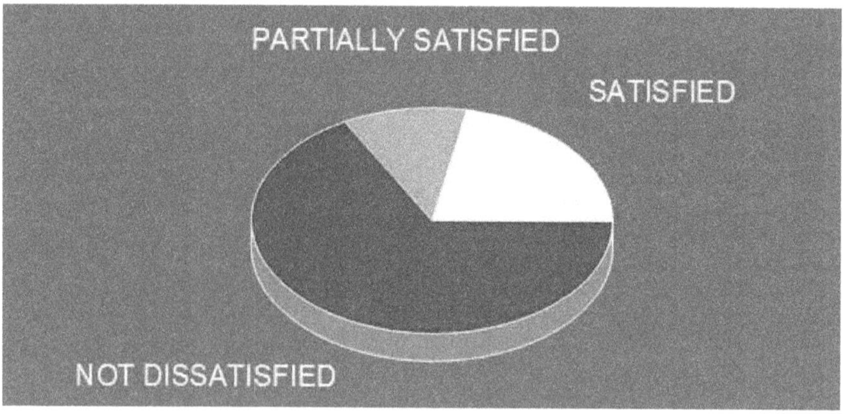

Figure F-10, Preventive Maintenance Programs

nothing more than lube routes and procedures. Therefore, once they have these developed, they stop. However, preventive maintenance encompasses much more than lubrication routes. Additionally, the lubrication position (often called an oiler or lubricator) is generally an entry level position. The lack of the technician's skills and training leads to substandard performance of the lubrication program. This results in excessive breakdowns, even though a company has a lubrication program. To be successful, maintenance organizations must go beyond the preliminaries and fully develop preventive maintenance programs.

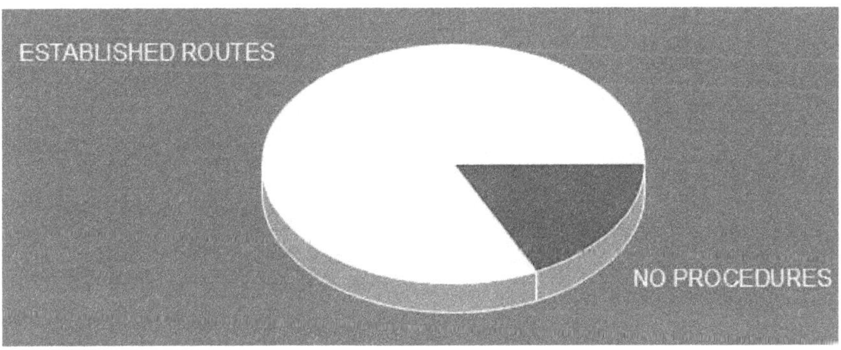

Figure F-11, Lubrication Routes

10. One final fact related to preventive maintenance is the lack of coordination between operations/facilities and maintenance. Almost three-fourths of all organizations experience problems in coordinating preventive maintenance with the operations/facilities group (see Figure F-12). The problem lies with communication. Either the maintenance organization has not communicated the need for the preventive maintenance or else the operations/facilities group is not educated properly about the benefits of preventive maintenance.

Good, credible communication must be established if preventive maintenance is to be effective.

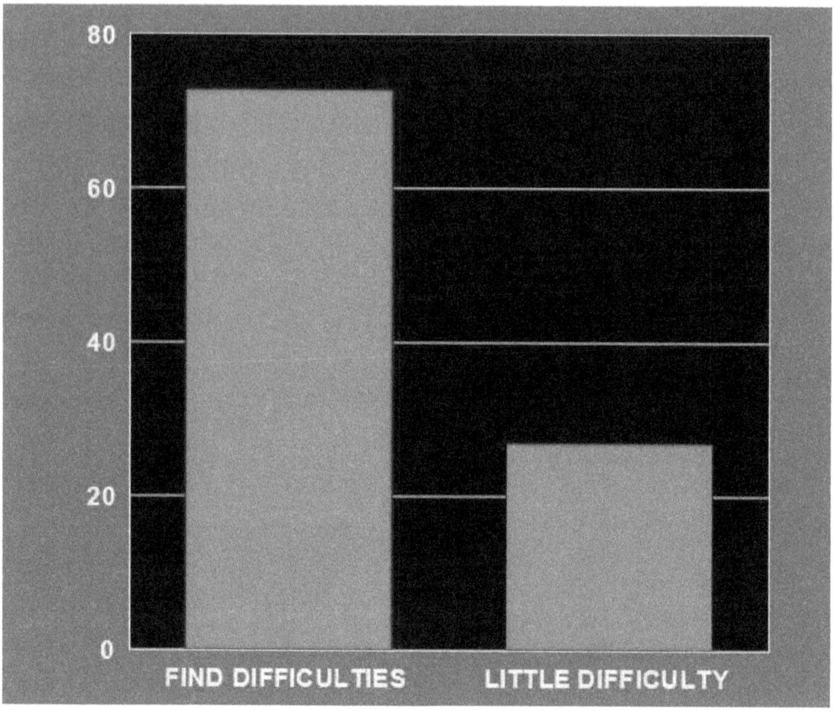

Figure F-12, Coordination Problems for Preventive Maintenance

11. Second only to maintenance labor, is the cost of maintenance materials. Depending on the type of operation/facility maintenance materials can range between 40–60% of the maintenance budget (see averages in Figure F-13). To successfully manage maintenance, materials must be given close scrutiny. The details of proper maintenance inventory management will be explained in Chapter 8.

12. Many companies try to remedy maintenance materials problems by overstocking the storeroom. This is a problem because most

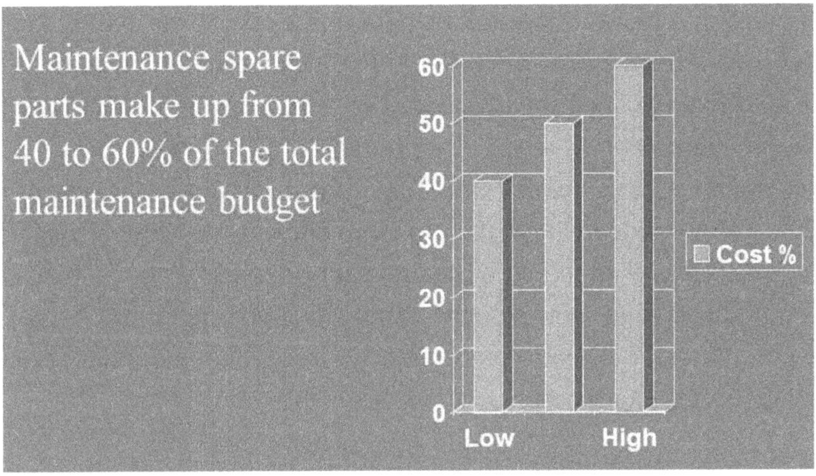

Figure F-13, Maintenance Spare Parts

will not take into account that inventory carrying costs are over 30% of the price of the items per year. For example, the cost of carrying a $1 million inventory is over $300,000 per year. It is for this reason the inventories must be kept low while still providing a satisfactory level of service and avoiding stock outs.

13. A third point of concern for maintenance materials is that maintenance is only responsible for their inventory in about 50% of the organizations. This means the other 50% of the time, someone else is telling maintenance what they should stock and how many they can stock or how many they can issue. It is the classic story of a support function telling its customer how to run their business.

14. While most will agree that maintenance costs are high, they don't know how high they are for their own site. In most cases, the costs of maintenance repairs are calculated as the cost of maintenance labor and the maintenance materials to effect the repair. However, that calculation leaves out the cost of lost production. The range for

this cost may be from two to fifteen times the cost of the maintenance repair as illustrated in Figure F-14. The average ratio is usually 4:1. Thus, a maintenance repair in labor and materials may be $10,000, but the actual cost is really closer to $50,000. Is it any wonder that maintenance costs are coming under closer scrutiny and control? In fact, it is estimated that in the next several years the annual maintenance costs will exceed the amount spent on yearly new capital investment. If one realizes the scrutiny given to yearly capital expenditures, is it any wonder that maintenance expenditures are being closely examined?

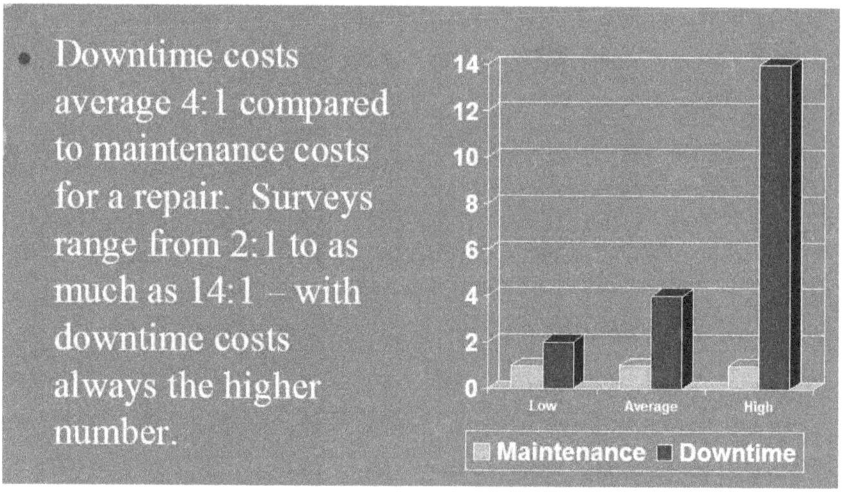

Figure 5-14, Maintenance Costs Compared to Downtime Costs

This Foreword has described the general condition of the maintenance function in facilities and industry today. Chapter 1 contains an analysis designed to look at specifics for each company. More detailed than the Foreword, it will enable a company to closely scrutinize its maintenance / asset management function.

Chapter 1

Analyzing Maintenance Management

Part 1 Maintenance Organizations

1. Maintenance organizational chart

A. Current and complete — 4 pts
B. Not complete or over 1 year old — 3 pts
C. Not current and not complete — 2 pts
D. None — 0 pts

2. Detailed job descriptions are available for

A. All maintenance positions (including supervisors) — 4 pts
B. All maintenance positions (except supervisors) — 3 pts
C. All maintenance supervisors (and no others) — 2 pts
D. Less than 50% of all maintenance positions — 1 pt
E. No job descriptions — 0 points

3. Maintenance supervisor to hourly maintenance employee ratio

A. 8:1 to 12:1 — 4 pts
B. 12:1 to 16:1 — 3 pts
C. Less than 8:1 — 2 pts
D. More than 16:1 — 1 pt
E. No shift maintenance supervisor — 0 pts

4. Maintenance planner to hourly maintenance employee ratio

A. 15:1 to 20:1 — 4 pts
B. 10:1 to 15:1 — 3 pts
C. 20:1 to 25:1 — 2 pts
D. 25:1 to 30:1 — 1 pt
E. No planner or anything other than above — 0 pts

5. Maintenance organizational assignments

A. Responsibilities fully documented — 4 pts
B. Responsibilities clear, good coverage, good dispatching — 3 pts
C. Informal supervision and coordination, some gaps in job coverage — 2 pts
D. Maintenance reports to production/operation — 1 pt
E. Unclear lines of authority, jurisdictional — 0 pts

6. Maintenance organization effort and attitude

A. Excellent, pride in workmanship at all levels — 4 pts
B. Steady work rate, professional operation — 3 pts
C. Average work pace, only a few complaints — 2 pts
D. Only occasional good efforts, frequent job delays, many complaints — 1 pt
E. Constant disagreements within maintenance organiza tion and between maintenance and operations / production/facilities — 0 pts

7. Maintenance shop/work area locations

A. Perfect — 4 pts
B. Good (some improvement possible) — 3 pts
C. Fair (major improvement possible) — 2 pts
D. Poor (major improvement required) — 1 pt
E. Unsuitable or nonexistent — 0 pts

8. Maintenance shop/work area layouts

A. Perfect — 4 pts
B. Good (some improvement possible) — 3 pts
C. Fair (major improvement possible) — 2 pts
D. Poor (major improvement required) — 1 pt
E. Unsuitable or nonexistent — 0 pts

9. Maintenance tools/ equipment quality and quantity

A. Perfect — 4 pts
B. Good (some improvement possible) — 3 pts
C. Fair (major improvement possible) — 2 pts
D. Poor (major improvement required) — 1 pt
E. Unsuitable or nonexistent — 0 pts

10. What percent of maintenance personnel are tied to a pay incentive plan based on output?

A. All — 4 pts
B. Greater than 90% — 3 pts
C. Greater than 75% — 2 pts
D. Greater than 50% — 1 pt
E. Less than 50% or other — 0 pts

Part 2 Training Programs in Maintenance

1. Supervisory training

A. All are trained when salaried and additional training is mandatory on a scheduled basis — 4 pts

B. All are trained when salaried and additional training is offered on an optional basis. — 3 pts

C. The majority are trained when salaried. — 2 pts

D. The majority are offered and attend training offered on an infrequent or irregular basis — 1 pt

E. Few are given initial training and little or no additional training is provided. — 0 pts

2. Planner training

A. All planners/schedulers have been to one or more public seminars providing instruction on maintenance planning and scheduling. — 4 pts

B. All planners/schedulers are provided with a written training program for maintenance planning. — 3 pts

C. All planners/schedulers receive 1 on 1 on the job training for at least 1 month. — 2 pts

D. Planner/scheduler training is on the job only. — 1 pt

E. There is no planner/scheduler training program. — 0 pts

3. Details of planner training subjects (Add 1 pt for each of the subjects covered; add 0 points if there is no planner training program.)

A. Work Order planning and execution

B. Material planning

C. Scheduling practices

D. Project planning

4. General quality and productivity training

A. Includes upper management, line supervision, hourly worker, support personnel — 4 pts

B. Includes upper management, line supervision, and hourly workers — 3 pts

C. Includes upper management, line supervision — 2 pts

D. Is for only upper management — 1 pt

E. No training program — 0 pts

5. Maintenance Craft training

a. Training is tied to a pay and progression program. — 4 pts

b. Formal job experience is required before employment and on the job training is provided. — 3 pts

c. Formal job experience is required before hire. — 2 pts

d. Training is provided by on the job experience after hire. — 1 pt

e. There are no formal training requirements for hire and no subsequent training is provided. — 0 pts

6. Maintenance training Intervals. Formal maintenance training is provided to ALL maintenance craft employees at the frequency of:

A. less than 1 year — 4 pts

B. between 12 to 18 months — 3 pts

C. between 18 to 24 months — 2pts

D. not to all employees, but to some in any of the above frequencies — 1 pt

E. no training is offered — 0 pts

7. Format of maintenance training

a. Training is a mix of classroom and lab exercises. — 4 pts
b. Training is all classroom. — 3 pts
c. Training is all in lab or workshop environment. — 2 pts
d. Training is all on the job. — 1 pt
e. No formal craft training program exists. — 0 pts

8. Training program instructors

A. Training is run by outside contract expert. — 4 pts
B. Training is run by staff subject expert. — 3 pts
C. Training is run by supervisors. — 2 pts
D. Training is run by hourly workers. — 1 pt
E. Training program does not exist. — 0 pts

9. The quality and skill level of the maintenance work force is

A. Perfect — 4 pts
B. Good (some improvement possible) — 3 pts
C. Fair (major improvement possible) — 2 pts
D. Poor (major improvement required) — 1 pt
E. Unsuitable — 0 pts

10. The quality and skill level of the supervisory group is

A. Perfect — 4 pts
B. Good (some improvement possible) — 3 pts
C. Fair (major improvement possible) — 2 pts
D. Poor (major improvement required) — 1 pt
E. Unsuitable — 0 pts

Part 3 Maintenance Work Orders

1. What percent of maintenance man-hours are reported to a work order?

A. 100% — 4 pts
B. 75% — 3 pts
C. 50% — 2 pts
D. 25% — 1 pt
E. less than 25% — 0 pts

2. What percent of maintenance materials are charged against a work order number whenissued?

A. 100% — 4 pts
B. 75% — 3 pts
C. 50% — 2 pts
D. 25% — 1 pt
E. less than 25% — 0 pts

3. What percent of total jobs performed by maintenance are covered by work orders?

A. 100% — 4 pts
B. 75% — 3 pts
C. 50% — 2 pts
D. 25% — 1 pt
E. less than 25% — 0 pts

4. What percent of the work orders processed in the system (either manual or computerized) are tied to an equipment/asset number?

A. 100% — 4 pts
B. 75% — 3 pts

C. 50% — 2 pts
D. 25% — 1 pt
E. less than 25% — 0 pts

5. What percent of the work orders are opened under a priority that would be identified as emergency, breakdown, or urgent?

A. Less than 20% — 4 pts
B. 20 to 29% — 3 pts
C. 30 to 39% — 2 pts
D. Greater than 39% — 0 pts

6. What percent of the work orders are available for historical data analysis?

A. 100% — 4 pts
B. 75% — 3 pts
C. 50% — 2 pts
D. 25% — 1 pt
E. less than 25% — 0 pts

7. What percent of the work orders are checked by a qualified individual for work quality and completeness?

A. 100% — 4 pts
B. 75% — 3 pts
C. 50% — 2 pts
D. 25% — 1 pt
E. less than 25% — 0 pts

8. What percent of the work orders are closed within 8 weeks from the date requested?

A. 100% — 4 pts
B. 75% — 3 pts
C. 50% — 2 pts
D. 25% — 1 pt
E. less than 25% — 0 pts

9. What percent of the work orders are generated from the preventive maintenance inspections?

A. 80–100% — 4 pts
B. 60–80% — 3 pts
C. 40–60% — 2 pts
D. 20–40% — 1 pt
E. less than 20% — 0 pts

10. Add one point for each of the categories you track for work orders:

A. Required downtime
B. Required craft hours
C. Required materials
D. Requestor's name

Part 4 Maintenance Planning and Scheduling

1. What percent of non-emergency work orders are completed within four weeks of the initial request?

A. More than 90% — 4 pts
B. Between 75 – 90% — 3 pts
C. Between 60 – 75% — 2 pts

D. Between 40 – 75% — 1 pt
E. Less than 40% — 0 pts

2. Work order planning (Add 1 point for each of the following.)

A. Craft required
B. Materials required
C. Tools required
D. Specific job instructions or job plan

3. Percentage of planned work orders experiencing delays due to poor or incomplete plans

A. Less than 10% — 4 pts
B. Between 10 – 20% — 3 pts
C. Between 20 – 40% — 2 pts
D. Between 40 – 50% — 1 pt
E. More than 50% — 0 pts

4. Who is responsible for planning the work orders?

A. A dedicated maintenance planner — 4 pts
B. A maintenance supervisor —2 pts
C. Each craft technician — 0 pts
D. Anyone else — 0 pts

5. Maintenance job schedules are issued

A. Weekly — 4 pts
B. Bi-weekly — 3 pts
C. Between 3 and 7 days — 2 pts
D. Daily — 1 pt
E. Any other frequency — 0 pts

6. The maintenance and production/ facilities scheduling meeting is held

A. Weekly — 4 pts
B. Bi-weekly — 3 pts
C. Between 3 and 7 days — 2 pts
D. Daily — 1 pt
E. Any other frequency — 0 pts

7. The backlog of maintenance work is available by (Add one point for each category.)

A. Craft required
B. Department/area requesting
C. Requestor
D. Date needed by

8. When the job is completed, the actual time, material, downtime, and other information is reported by

A. The craft technician performing the work — 4 pts
B. The supervisor of the group — 3 pts
C. Anyone else — 2 pts
D. Information is not recorded — 0 pts

9. What percent of the time are the actuals are compared to the estimates for monitoring planning effectiveness?

A. More than 90% — 4 pts
B. Between 75 – 90% — 3 pts
C. Between 60 – 75% — 2 pts
D. Between 40 – 75% — 1 pt
E. Less than 40% — 0 pts

10. What is the reporting relationship between planners and supervisors?

A. Both report to the same maintenance manager. — 4 pts
B. The planner reports to the supervisor. — 2 pts
C. The supervisor reports to the planner. — 2 pts
D. The supervisor and planner report to operations/facilities. — 0 pts

Part 5 Preventive Maintenance

1. The preventive maintenance program includes (Add 1 point for each type included.)

A. Lubrication checklists
B. Detailed inspection checklists
C. Personnel assigned specifically assigned to the PM program
D. Predictive maintenance diagnostics such as vibration analysis, oil sample analysis,infrared heatmonitors,etc.

2. What percent of the PM inspection/task checklists are checked to insure completeness?

A. More than 90% — 4 pts
B. Between 75 – 90% — 3 pts
C. Between 60 – 75% — 2 pts
D. Between 40 – 75% — 1 pt
E. Less than 40% — 0 pts

3. What percent of the plant critical equipment is covered by a preventive maintenance program?

A. More than 90% — 4 pts
B. Between 75 – 90% — 3 pts

C. Between 60 – 75% — 2 pts
D. Between 40 – 75% — 1 pt
E. Less than 40% — 0 pts

4. What percent of the PM program is checked annually against an equipment item's history to insure good coverage?

A. More than 90% — 4 pts
B. Between 75 – 90% — 3 pts
C. Between 60 – 75% — 2 pts
D. Between 40 – 75% — 1 pt
E. Less than 40% — 0 pts

5. What percent of the PMs are completed within 1 week of the due date?

A. More than 90% — 4 pts
B. Between 75 – 90% — 3 pts
C. Between 60 – 75% — 2 pts
D. Between 40 – 75% — 1 pt
E. Less than 40% — 0 pts

6. What determines the frequency of a PM inspection or task/service interval?

A. The program is condition-based. — 4 pts
B. The program is based on a combination of equipment run time and fixed calendar interval. — 3 pts
C. The program is based on equipment run time only. — 2 pts
D. The program is based on calendar intervals. — 1 pt
E. The program is dynamic and is scheduled based on completion date of previous task. — 0 pts

7. What percent of the inspections/tasks include safety information, detailed inspection instructions, material requirements, and labor estimates?

A. More than 90% — 4 pts
B. Between 75 – 90% — 3 pts
C. Between 60 – 75% — 2 pts
D. Between 40 – 75% — 1 pt
E. Less than 40% — 0 pts

8. What percent of the corrective action work orders are generated from the PM inspection program?

A. More than 90% — 4 pts
B. Between 75 – 90% — 3 pts
C. Between 60 – 75% — 2 pts
D. Between 40 – 75% — 1 pt
E. Less than 40% — 0 pts

9. On what percent of the program are PM actuals and results checked annually for time and material estimate accuracy?

A. More than 90% — 4 pts
B. Between 75 – 90% — 3 pts
C. Between 60 – 75% — 2 pts
D. Between 40 – 75% — 1 pt
E. Less than 40% — 0 pts

10. Who is responsible for performing preventive maintenance tasks?

A. Dedicated preventive maintenance personnel — 4 pts
B. Specific individuals on each crew — 3 pts
C. Any individual on a crew — 2 pts

D. Entry level craftworkers — 1 pt
E. Operating personnel — 0 pts

Part 6 Maintenance Inventory and Purchasing

1. What percent of the time are materials in stores when required by the maintenance organization?

A. More than 95% — 4 pts
B. Between 80 – 95% — 3 pts
C. Between 70 – 80% — 2 pts
D. Between 50 – 70% — 1 pt
E. Less than 50% — 0 pts

2. What percent of the items in inventory appear in the maintenance stores catalog?

A. More than 90% — 4 pts
B. Between 75 – 90% — 3 pts
C. Between 60 – 75% — 2 pts
D. Between 40 – 75% — 1 pt
E. Less than 40% — 0 pts

3. Who controls what are stocked as maintenance inventory items?

A. Maintenance — 4 pts
B. Anyone else — 0 pts

4. The maintenance stores catalog is produced by

A. Alphabetic and numeric listings — 4 pts
B. Alphabetic only — 2 pts
C. Numeric only — 2 pts
D. Catalog is incomplete or non-existent — 0 pts

5. The aisle/bin location is specified for what percent of the stores items?

A. More than 95% — 4 pts
B. Between 90 – 95% — 3 pts
C. Between 80 – 90% — 2 pts
D. Between 70 – 80% — 1 pt
E. Less than 70% — 0 pts

6. What percent of the maintenance stores items are issued to a work order or account number upon leaving the store?

A. More than 95% — 4 pts
B. Between 90 – 95% — 3 pts
C. Between 80 – 90% — 2 pts
D. Between 70 – 80% — 1 pt
E. Less than 70% — 0 pts

7. For what percent of the inventory are maximum and minimum levels for the maintenance stores items specified?

A. More than 95% — 4 pts
B. Between 90 – 95% — 3 pts
C. Between 80 – 90% — 2 pts
D. Between 70 – 80% — 1 pt
E. Less than 70% — 0 pts

8. What percent of the critical maintenance materialis stocked in the warehouse or in a location readily accessible when the material is required?

A. More than 95% — 4 pts
B. Between 90 – 95% — 3 pts
C. Between 80 – 90% — 2 pts

D. Between 70 – 80% — 1 pt
E. Less than 70% — 0 pts

9. What percent of the time are maintenance stores inventory levels daily updated upon receipt of materials?

A. More than 95% — 4 pts
B. Between 90 95% — 3 pts
C. Between 80 – 90% — 2 pts
D. Between 70 – 80% — 1 pt
E. Less than 70% — 0 pts

10. What percent of the items are checked for at least one issue every six months?

A. More than 90% — 4 pts
B. Between 80 – 90% — 3 pts
C. Between 70 – 80% — 2 pts
D. Between 50 – 70% — 1 pt
E. Less than 50% — 0 pts

Part 7 Maintenance Automation

1. What percentage of all maintenance operations utilizes a CMMS?

A. More than 90% — 4 pts
B. Between 75 – 90% — 3 pts
C. Between 60 – 75% — 2 pts
D. Between 40 – 75% — 1 pt
E. Less than 40% — 0 pts

2. What percentage of maintenance activities are planned and scheduled through the CMMS?

A. More than 90% — 4 pts
B. Between 75 – 90% — 3 pts
C. Between 60 – 75% — 2 pts
D. Between 40 – 75% — 1 pt
E. Less than 40% — 0 pts

3. What percentage of the maintenance inventory and purchasing functions are performed in the system?

A. More than 90% — 4 pts
B. Between 75 – 90% — 3 pts
C. Between 60 – 75% — 2 pts
D. Between 40 – 75% — 1 pt
E. Less than 40% — 0 pts

4. Is the CMMS and the production scheduling system

A. Integrated — 4 pts
B. Interfaced — 3 pts
C. There is no connection. — 0 pts

5. Are the CMMS and the payroll/ timekeeping system

A. Integrated — 4 pts
B. Interfaced — 3 pts
C. There is no connection. — 0 pts

6. Are the CMMS and the financial/accounting system

A. Integrated — 4 pts
B. Interfaced — 3 pts
C. There is no connection. — 0 pts

7. What percent of the maintenance personnel are using the system for their job functions with a high level of proficiency?

A. More than 90% — 4 pts
B. Between 75 – 90% — 3 pts
C. Between 60 – 75% — 2 pts
D. Between 40 – 59% — 1 pt
E. Less than 40% — 0 pts

8. CMMS data are structured and maintained to facilitate reporting

A. More than 90% — 4 pts
B. Between 75 – 90% — 3 pts
C. Between 60 – 75% — 2 pts
D. Between 40 – 59% — 1 pt
E. Less than 40% — 0 pts

9. Are CMMS data utilized on a regular basis to make cost effective management decisions?

A. Yes — 4 pts
B. Sometimes — 2 pts
C. No — 0 pts

10. Are CMMS data used to verify progressive ROI?

A. Yes — 4 pts
B. No — 0 pts

Part 8 Operations/Facilities Involvement

1. What percent of Operations personnel generate work order requests?

A. More than 90% — 4 pts
B. Between 75 – 90% — 3 pts
C. Between 60 – 75% —2 pts
D. Between 40 – 60% —1 pt
E. Less than 40% — 0 pts

2. What percent of Facilities personnel generate work order requests?

A. More than 90% — 4 pts
B. Between 75 – 90% — 3 pts
C. Between 60 – 75% — 2 pts
D. Between 40 – 60% — 1 pt
E. Less than 40% — 0 pts

3. Operations work order priority is set for maintenance

A. Weekly in a joint Operations / Maintenance meeting — 4 pts
B. Daily in a joint Operations / Maintenance meeting — 2 pts
C. It is set by Maintenance with minimal Operations input. —1 pt
D. It is random and based on emergency needs. — 0 pts

4. Facility work order priority is set for maintenance

A. Weekly in a joint Facility / Maintenance meeting —4 pts
B. Daily in a joint Facility / Maintenance meeting —2 pts

C. It is set by Maintenance with minimal Facilities input. —1 pt

D. It is random and based on emergency needs. — 0 pts

5. Are Operations/Operators responsible and involved in the upkeep and performance of assets?

A. Yes 4 pts

B. No — 0 pts

6. Add 1 point for each task that operators are trained and certified to perform.

A. Inspections —1 pt

B. Lubrication —1 pt

C. Minor maintenance task — 1 pt

D. Assist in maintenance repair work —1 pt

7. Operators follow-up and have a sign-off on completed work orders

A. 90% of the time—- 4 pts

B. 75 – 90% of the time — 3 pts

C. 60 – 75% of the time — 2 pts

D. 40 – 60% of the time — 1 pts

E. Less than 40% of the time — 0 pts

8. Facilities personnel follow-up and have a sign-off on completed work orders

A. 90% of the time — 4 pts

B. 75 – 90% of the time — add 3 pts

C. 60 – 75% of the time — add 2 pts

D. 40 – 60% of the time — add 1 pt

E. Less than 40% of the time — 0 pts

9. Maintenance is included in production/process scheduling meetings

A. All of the time —4 pts
B. Most of the time —3 pts
C. Occasionally — 2 pts
D. Seldom —1 pt
E. Never — 0 pts

10. Does asset-focused communication exists among maintenance, operations, engineering, and facilities personnel?

A. All of the time — 4 pts
B. Most of the time — 3 pts
C. Occasionally — 2 pts
D. Seldom — 1 pt
E. Never — 0 pts

Part 9 Maintenance Reporting

1. What percent of the time are the maintenance reports distributed / available on a timely basis to the appropriate personnel?

A. More than 90% — 4 pts
B. Between 75 – 90% — 3 pts
C. Between 60 – 75% — 2 pts
D. Between 40 – 75% — 1 pt
E. Less than 40% — 0 pts

2. What percent of the time are the reports distributed / available within 1 day of the end of the time period specified in the report?

A. More than 90% — 4 pts
B. Between 75 – 90% — 3 pts
C. Between 60 – 75% — 2 pts
D. Between 40 – 75% — 1 pt
E. Less than 40% — 0 pts

3. Add one point for each of the following equipment reports you produce.

A. Equipment downtime in order of highest to lowest total hours (weekly or monthly)
B. Equipment downtime in order of highest to lowest in total lost production dollars (weekly or monthly)
C. Maintenance cost for equipment in order of highest to lowest (weekly or monthly)
D. MTTR and MTBF for equipment

4. Add one point for each of the following preventive maintenance reports you produce.

A. PM overdue in order of oldest to most recent
B. PM cost per equipment item in descending order
C. PM hours versus total maintenance hours per item, expressed as a percentage
D. PM costs versus total maintenance costs per equipment item, expressed as a percentage

5. Add one point for each of the personnel reports you produce.

A. Time report showing hours worked by employee divided by work order

B. Time report showing hours worked by craft in each department/area

C. Time report showing total hours spent by craft on emergency / preventive/ normal work

D. Time report showing total overtime hours compared to regular hours

6. Add one point for each of the planning reports you produce.

A. Total work order costs estimated versus total work order actual costs by individual work order, by super visor, or by craft

B. A backlog report showing the total hours ready to schedule versus the craft capacityper week

C. A planning efficiency report showing the hours and materials planned versus the actual hours and materials used per work order

D. A planning effectiveness report showing the number of jobs closed out that were 20% over or under the labor or material estimates by planner and supervisor

7. Add one point for each of the scheduling reports you produce.

A. Hours worked as scheduled compared to actual hours worked

B. Weekly crew or craft capacity averaged for the last 20 weeks

C. Total number of maintenance work orders scheduled compared to the actual number of work orders completed

D. Number of work orders spent on preventive maintenance compared to emergency mainte nance and normal maintenance

8. Add one point for each of the Inventory reports you produce.

A. Stock catalog by alphabetical and numerical listing

B. Inventory valuation report

C. Inventory performance report showing stock outs, level of service, turnover rate, etc.

D. Inventory where used report

9. Add one point for each of the Purchasing reports you produce.

A. Vendor performance showing promised and actual delivery dates

B. Price performance, showing the quoted and actual prices

C. Buyer or purchasing agent performance report

D. Non-stock report showing all direct buys for items not carried in stock for a specified period

10. Add one point for each administrative report you produce.

A. Monthly maintenance costs versus monthly maintenance budget with a year to date total

B. Comparison of labor and material costs as a percentage of total maintenance costs

C. Total costs of outside contractor usage broken down by contractor / project

D. Maintenance cost per unit of production (or by square foot for facilities)

Part 10 Predictive Maintenance

1. The predictive maintenance program includes vibration analysis for

A. Critical/Non-critical assets — 4 pts

B. Critical only — 2 pts

C. None — 0 pts

2. The predictive maintenance program includes thermography for

A. Critical/Non-critical assets — 4 pts

B. Critical only — 2 pts

C. None — 0 pts

3. The predictive maintenance program includes oil analysis for

A. Critical/Non-critical assets — 4 pts

B. Critical only — 2 pts

C. None — 0 pts

4. The predictive maintenance program utilizes sonic/ ultrasound techniques for

A. Critical/Non-critical assets — 4 pts

B. Critical only — 2 pts

C. None — 0 pts

5. CONDITION BASED MONITORING IS INCLUDED IN THE PREDICTIVE MAINTENANCE PROGRAM.

A. Yes — 4 pts
B. No — 0 pts

6. ARE THE PREDICTIVE MAINTENANCE SYSTEM AND THE CMMS

A. Integrated — 4 pts
B. Interfaced — 3 pts
C. Not electronically connected — 0 pts

7. IS THE PREDICTIVE MAINTENANCE DATA USED TO GENERATE PREVENTIVE / CORRECTIVE MAINTENANCE WORK ORDERS?

A. Yes — 4 pts
B. No — 0 pts

8. PERSONNEL ARE EXCLUSIVELY ASSIGNED TO THE PREDICTIVE MAINTENANCE PROGRAM.

A. Yes — 4 pts
B. No — 0 pts

9. PREDICTIVE WORK IS INCLUDED AS PART OF THE WEEKLY WORK SCHEDULE.

A. Yes — 4 pts
B. No — 0 pts

10. IS PREDICTIVE MAINTENANCE DATA USED TO IMPROVE ASSET PERFORMANCE AND LIFE EXPECTANCY?

A. Yes — 4 pts
B. No — 0 pts

Part 11 Reliability Engineering

1. The organization has a reliability engineering attitude/mentality.

A. Yes, it exists as part of the work culture.— 4 pts
B. Yes, it exists and is being developed as part of the work culture. — 3 pts
C. No — 0 pts

2. Complete and accurate asset data is available for Reliability Centered Maintenance analysis on

A. 90% of the assets — 4 pts
B. 75 – 90% of the assets — 3 pts
C. 60 – 75% of the assets — 2 pts
D. 40 – 60% of the assets — 1 pt
E. Less than 40% of the assets — 0 pts

3. Is the RCM methodology used to adjust/refine the PM/PDM program?

A. Yes — 4 pts
B. No — 0 pts

4. How often is RCM analysis conducted on all assets?

A. Annually — 4 pts
B. Every 6 months — 3 pts
C. Every 2 years — 2 pts
D. Every 3 years — 1 pt
E. Any longer — 0 pts

5. The work order history is accurate in tracking the causes of failures

A. 90% of the asset's history — 4 pts
B. 75 – 90% of the asset's history — 3 pts
C. 60 – 75% of the asset's history — 2 pts
D. 40 – 60% of the asset's history — 1 pt
E. Less than 40% of the asset's history — 0 pts

6. Are failures clearly identified?

A. 90% of the asset's history — 4 pts
B. 75 – 90% of the asset's history — 3 pts
C. 60 – 75% of the asset's history — 2 pts
D. 40 — 60% of the asset's history — 1 pt
E. Less than 40% of the asset's history — 0 pts

7. Failure analysis is conducted using analysis tools such as "5 whys," fishbone, Pareto, etc., to assure accuracy and consistency of the effort.

A. Yes — 4 pts
B. No — 0 pts

8. Dedicated personnel are permanently assigned to maintain the RCM program.

A. Yes — 4 pts
B. No — 0 pts

9. Management views RCM as a value added activity?

A. Yes — 4 pts
B. No — 0 pts

10. Are there methods in place for measuring the effectiveness of the reliability engineering effort?

A. Yes — 4 pts
B. No — 0 pts

Part 12 Maintenance — General Practices

1. Is the total organization focused on asset management / utilization / optimization?

A. Yes — 4 pts
B. No — 0 pts

2. The maintaining function is perceived as value add by: (Add 1 point for each area.)

A. Management — 1 pt
B. Operations — 1 pt
C. Maintenance — 1 pt
D. Stores and Purchasing — 1 pt

3. The maintenance data collection system (CMMS/EAM) is utilized by (Add 1 point for each area.)

A. Management — 1 pt
B. Operations — 1 pt
C. Maintenance — 1 pt
D. Stores and Purchasing — 1 pt

4. Operators are used for first line maintenance functions

A. In all areas — 4 pts
B. In some areas — 3 pts

C. In a few areas — 2 pts
D. Not used at all — 0 pts

5. Is the overall equipment effectiveness (OEE) calculated on key assets, processes, and facilities

A. > 90% — 4 pts
B. > 60% — 3 pts
C. > 30% — 2 pts
D. < 30% — 0 pts

6. Operational decisions are made taking into account equipment reliability / availability

A. > 90% of the time — 4 pts
B. > 60% of the time — 3 pts
C. > 30% of the time — 2 pts
D. < 30% of the time — 0 pts

7. Have the right "soft skills" training classes been conducted for appropriate personnel? (Examples of soft skills are communications, leadership, etc.)

A. > 90% of the personnel — 4 pts
B. > 60% of the personnel — 3 pts
C. > 30% of the personnel — 2 pts
D. < 30% of the personnel — 0 pts

8. Have the right "technical" training classes been conducted for appropriate personnel (e.g., maintenance and operations)?

A. > 90% of the personnel — 4 pts
B. > 60% of the personnel — 3 pts
C. > 30% of the personnel — 2 pts
D. < 30% of the personnel — 0 pts

9. Does the maintenance program comply with regulatory requirements and programs?

A. Yes — 4 pts

B. No — 0 pts

10. Are the financial effects of equipment availability / reliability understood and communicated to everyone?

A. Yes — 4 pts

B. No — 0 pts

Part 13 Financial Optimization

1. Downtime duration is consistently tracked

A. For all assets — 4 pts

B. For key assets only — 2 pts

C. Not tracked at all — 0 pts

2. Downtime cost is clearly identified for key assets, processes, and facilities

A. For all areas — 4 pts

B. For key areas only — 2 pts

C. Not tracked at all — 0 pts

3. Root causes of downtime are accurately and consistently tracked

A. For all assets — 4 pts

B. For key assets only — 2 pts

C. Not tracked at all — 0 pts

4. MAINTENANCE COSTS ARE CLEARLY AND ACCURATELY TRACKED

A. For all assets — 4 pts
B. For key assets only — 2 pts
C. Not tracked at all — 0 pts

5. WHAT PORTION OF OTHER CONTRIBUTING COSTS (ENERGY, QUALITY, CONTRACTORS) ARE AVAILABLE FOR ANALYSIS?

A. All the costs — 4 pts
B. Some of the costs — add 2 pts
C. None of the costs — 0 pts

6. ARE TOTAL LIFE CYCLE COSTS COMPARED WHEN MAKING DECISIONS?

A. All cost factors — 4 pts
B. Some cost factors — 2 pts
C. No cost factors — 0 pts

7. EFFICIENCY LOSS COSTS ARE AVAILABLE AND ACCURATE FOR

A. 90% of the assets — 4 pts
B. 60 – 90% of the assets — 3 pts
C. 40 – 60% of the assets — 2 pts
D. 30 – 39% of the assets — 1 pt
E. Less than 30% of the assets — 0 pts

8. IS A DEDICATED INDIVIDUAL OR TEAM ASSIGNED TO ANALYZE FINANCIAL COSTS?

A. Yes — 4 pts
B. No — 0 pts

9. ARE STORES AND PURCHASING COSTS ACCURATELY TRACKED?

A. Yes — 4 pts
B. No — 0 pts

10. How available is financial information?

A. Available on demand — 4 pts
B. Available daily — 3 pts
C. Available weekly — 2 pts
D. Available monthly — 1 pt
E. Not available — 0 pts

Part 14 Asset Care Continuous Improvement

1. Is there VISIBLE management support for continuous improvement efforts?

A. Strong support — 4 pts
B. Moderate support — 3 pts
C. Weak support — 2 pts
D. None — 0 pts

2. Does the organization support continuous improvement efforts?

A. Strong support — 4 pts
B. Moderate support — 3 pts
C. Weak support — 2 pts
D. None — 0 pts

3. If the company has recently downsized, has it affected the organization?

A. Positive effect — 4 pts
B. No effect — 2 pts
C. Adverse effect — 0 pts

4. Past support of improvement efforts has been

A. Excellent — 4 pts
B. Good — 2 pts
C. Poor — 0 pts

5. The spirit of cooperation between plant management and labor is

A. Excellent — 4 pts
B. Good — 2 pts
C. Poor — 0 pts

6. The following personnel are focused on continuous improvement

A. Management — 1 pt
B. Maintenance — 1 pt
C. Stores/Purchasing — 1 pt
D. All other personnel — 1 pt
E. No one — 0 pts

7. Does management support on-going training designed to enhance employee skills?

A. Yes — 4 pts
B. No — 0 pts

8. Do continuous improvement efforts focus on ROI?

A. Yes — 4 pts
B. No — 0 pts

9. ARE CONTINUOUS IMPROVEMENT EFFORTS TIED TO RELIABILITY ENGINEERING?

A. Yes — 4 pts
B. No – 0 pts

10. HOW OFTEN DO COMPETITIVE FORCES INFLUENCE CONTINUOUS IMPROVEMENT EFFORTS?

A. 90% of the time — 4 pts
B. 60 – 90% of the time — 3 pts
C. 40 – 60% of the time — 2 pts
D. 30 – 40% of the time — 1 pt
E. Less than 30% of the time — 0 pts

PART 15 MAINTENANCE CONTRACTING

1. THE CONTRACT REQUEST PROCESS INCLUDES (ADD 1 POINT FOR EACH ELEMENT INCLUDED.)

A. A formal process for requesting contract work
B. An established approval process based on dollar value
C. A mechanism to control what is and is not contracted
D. The request process is automated

2. THE RESPONSIBILITY FOR THE CONTRACTING OF WORK

A. Is handled by a contract specialist function solely focused on contracts — 4 pts
B. Is handled by a contract function but not solely focused on contracting — 3 pts
C. Is handled by various job functions who have contracting responsibility — 2 pts
D. Contracting is handled by whoever needs contract work — 0 pts

3. The approved contractors list includes (Add 1 point for each element included.)

A. A primary contractor for all required services

B. An alternate in the event the primary is unavailable

C. Validated requirements such as insurance, crafts rates, etc.

D. Phone numbers for all contacts

4. The owner provides the following level of supervision for contracted field execution

A. Dedicated company personnel to administer the contract in the field — 4 pts

B. Company personnel to administer the contracts, but it is not their primary function — 2 pts

C. There is no organizational structure to administer field execution — 0 pts

5. Contractor safety is

A. A joint effort (joint accountability) between the site and the contractors — 4 pts

B. Highly important and the contractors are held accountable — 3 points

C. Equally important with the execution of the work — 1 point

D. Not as important as work execution — 0 points

6. The computer systems exists with the following contracting functionality on the front end of the process

A. Contract requisition — 1 pt

B. Contract administration — 1 pt

C. Contract approval processing — 1 pt
D. Administration of contracts in the field; releasing of work — 1 pt
E. If there is no electronic contract system — 0 pts

7. The computer systems exist with the following contracting functionality on the back end of the process

A. Contract employees linked to contract (rates, skills, certifications, etc.) — 1 pt
B. Electronic time sheets, paperless system — 1 pt
C. Gate security electronically linked — 1 pt
D. Invoice payment processes — 1 pt
E. If there is no electronic contract system — 0 pts

8. The contracting systems is

A. Integrated with the site's CMMS — 4 pts
B. A fully functional system with interfaces to the site CMMS — 3 pts
C. A stand-alone system — 2 pts
D. There is no system — 0 pts

9. The relationship between the site personnel and the site contractors is

A. A partnership focused on efficient and effective work execution — 4 pts
B. A supportive–superior / subordinate relationship — 3 pts
C. Contractors are accepted but not part of the team effort — 2 pts
D. Adversarial; they are a necessary evil — 0 pts

10. The invoicing / cost tracking process is

A. Highly developed so that those responsible know the costs daily — 4 pts

B. Moderately developed; costs are available weekly — 3 pts

C. The costs are only known when invoices are processed (long lead time) — 2 pts

D. Costs are not readily available to those controlling the work — 0 pts

Part 16 Document Management

1. The site's document management system is

A. Electronic and fully integrated with other systems — 4 pts

B. Electronic and interfaced with other systems — 3 pts

C. Electronic but stand-alone — 2 pts

D. Paper – non electronic — 0 pts

2. The number of drawings included in the system is

A. 80% to 100% — 4 pts

B. 50% to 80% — 3 pts

C. 30% to 50% — 1 pt

D. less than 30% — 0 pts

3. The site's timing for migration to a fully functional and utilized document management systems is

A. 1 to 3 years — 4 pts

B. 3 to 5 years — 3 pts

C. 5 or more years — 1 pt

D. No plan — 0 pts

4. Document control procedures and associated work process

A. Exist for the system — 4 pts

B. Are under development — 2 pts

C. Do not exist — 0 pts

5. The personnel at the site

A. Have received training; fully understand and use the document management procedures — 4 pts

B. Are in the process of being trained to use the procedures — 3 pts

C. Do not follow the procedures or they donot exist — 0 pts

D. Add 1 point if procedure refresher training is provided periodically

6. The system has detailed indexing and search capabilities that

A. Make documents simple and easy to find — 4 pts

B. Make documents difficult to find — 2 pts

C. Indexing and search capability does not exist — 0 pts

7. The level of accessibility for users is

A. Users have access and it is easy to obtain information — 4 pts

B. Users have access, but it is hard to find information — 3 pts

C. Users must go to a separate group who provides the information — 1 pt

D. Information is not accessible for most users — 0 pts

8. The quality and level of document version control is

A. Excellent — 4 pts
B. Good — 3 pts
C. Average — pts
D. Poor — 1 pt
E. Very Poor — 0 pts

9. The number of documents included in the system is

A. 80% to 100% — 4 pts
B. 50% to 80% — 3 pts
C. 30% to 50% — 1 pt
D. less than 30% — 0 pts

10. The level of accessibility for users is

A. Users have access and it is easy to obtain information — 4 pts
B. Users have access but it is hard to find information — 3 pts
C. Users must go to a separate group who provides the information — 1 pt
D. Information is not accessible for most users 0 pts

The Results

After completing the survey above, an organization should have insight into the effectiveness of their maintenance/ asset management function. However, the survey total is a possible 640 points. What is considered good and what is considered poor for the survey? During the last 15 years, over 1200 companies have participated by measuring themselves against the criteria in this survey. The representative results are detailed below. It will be useful to examine the results, looking for maintenance process areas that could use improvement.

Although these are the individual totals, it is interesting to plot the results in a spider diagram using a product like Microsoft's Excel. The results for the representative survey are shown in Figure 1-1.

Using this survey, a modified version, or any other similar survey is an important prerequisite to beginning a benchmarking project. This analysis tool or a similar one is necessary for conducting an internal analysis. It will allow an organization to identify its starting point. How to use this information in a benchmarking project will be explained in Chapter 2.

Part 1	Maintenance Organizations	24.5
Part 2	Training Programs in Maintenance	22.5
Part 3	Maintenance Work Orders	26.3
Part 4	Maintenance Planning and Scheduling	22.5
Part 5	Preventive Maintenance	24.1
Part 6	Maintenance Inventory and Purchasing	28.4
Part 7	Maintenance Automation	22.0
Part 8	Operations/Facilities Involvement	22.2
Part 9	Maintenance Reporting	13.4
Part 10	Predictive Maintenance	20.0
Part 11	Reliability Engineering	15.7
Part 12	Maintenance — General Practices	29.5
Part 13	Financial Optimization	23.9
Part 14	Asset Care Continuous Improvement	26.9
Part 15	Maintenance Contracting	21.3
Part 16	Document Management	23.1
Total		366.0

Figure 1-1, Sample Spider Diagram

Chapter 2

Benchmarking Fundamentals

Benchmarking. Benchmarks. Best Practices. All of these terms are used in business today. But are they just buzzwords, or do the words have real meaning? Are they useful tools that can be used to improve business practices today? Let's begin with some definitions.

Definitions

Benchmarking

There are several differing definitions for benchmarking. For example, Xerox Corporation defines benchmarking as follows:

> *The search for industry Best Practices*
> *which lead to Superior Performance.*

The key elements in this definition are the terms Best Practices and Superior Performance. What is a Best Practice? It is a series of processes that enables a company to become a leader in its respective marketplace. However, Best Practices are not the same for all companies. If a company is in a declining market, where the pressures are to maximize profits with a fixed sales volume, one set of Best Practices might allow market leadership. If, however, the company is in a growth mode with profits dictated by gaining rapid market share, another set of Best Practices allows market leadership. Therefore, Best is determined by business conditions, not by a fixed set of business practices.

A note on the term *Best Practices*. To avoid Best Practices just becoming a buzzword in many organizations, the words have been changed to Better Practices. Because no universally accepted Best Practices exist, Better Practices can be utilized in organizations with a continuous improvement business philosophy.

The second key phrase in the Xerox definition is *Superior Performance*. Many companies use benchmarking to be "as-good-as" their competitors. However, there is very little that can be gained if the goal of benchmarking is merely to achieve status quo. Benchmarking is a continuous improvement tool; it should be utilized by companies that are striving to achieve superior performance in their respective marketplace. Only then can benchmarking be properly utilized as a true continuous improvement tool.

An alternative definition for benchmarking proposed by Robert Camp in *The Benchmarking Handbook*, is:

An ongoing process of measuring and improving business practices against the companies that can be identified as the best worldwide

A key point in this definition is the idea of improving, not maintaining status quo. A second point is the worldwide search for best companies. Most companies have international competitors. It would be naive to think that Best Practices are limited to one country or one geographical location. Logically, information that allows companies to improve their competitive positions must be gathered from Best companies, no matter where they are geographically located.

Companies striving to improve must not accept past constraints, especially the "not invented here" paradigm. Companies that fail to develop a global perspective will soon be replaced by competitors who had the insight to think globally. In order to make rapid continuous improvement, companies must be able to think outside the box, that is, to examine their business from external perspectives. The more innovative the ideas that are discovered, the greater the potential rewards that can be gained from the adaptation of the ideas.

A third perspective on benchmarking states:

Benchmarking sources "Best Practices"
to feed continuous improvement.

This statement adds another dimension to benchmarking, that of having an external perspective. Research shows that major innovations in any business sector comes from an external market sector and is adapted to improve the practices of the company. Because "Best Practices" can be found in any industry section, finding them requires companies to look outside their industrial vertical market. In today's competitive business environment, the need to develop this external perspective is critical to survival.

A final perspective says that benchmarking is the process of continuously comparing and measuring an organization with business leaders anywhere in the world to gain information that will help the organization take action to improve its performance. The common thread of studying other companies to gain information that allows the company to become more competitive is clear throughout all four descriptions. Unless a company clearly understands the processes and procedures that allow it to become the best, little value is derived from benchmarking.

Competitive Analysis

There is considerable confusion between the terms benchmarking and competitive analysis. Whereas benchmarking references external business sectors for information, competitive analysis shows how firms compare only with their competitors. A competitive analysis produces a ranking in comparison to direct competitors; it does not show how to improve business processes. This is the major difference between benchmarking and a competitive analysis.

Benchmarking, compared with a competitive analysis, provides a deep understanding of the processes and skills that create superior performance. Unless the deep understanding is achieved, little benefit is derived from benchmarking. The question should be asked "What is your Goal for benchmarking?" If the goal is only to meet some industry standard, then there is little to gain from investing in benchmarking to be superior.

A competitive analysis is unlikely to lead to or highlight significant breakthrough opportunities that could change long entrenched paradigms of any particular market segment. This is because business paradigms are similar for look-alike businesses in similar markets. Whereas a competitive analysis can result in incremental improvements for a business, breakthrough strategies are derived from taking an external perspective. By discovering and understanding external best practices, quantum leaps in business process improvements can be derived.

During the past decades, competitive analyses have helped companies improve their respective market positions. Benchmarking takes over where a competitive analysis ends. Benchmarking will allow companies to move from a parity business position to a superiority position. Lessons learned from the best companies can help any company.

One of the major differences between benchmarking and a competitive analysis is the level of documentation. A competitive analysis focuses on meeting some type of published industry standard. Therefore, all that may be required is some published number. By comparison, benchmarking focuses not on a number, but on the process that allows such a standard not only to be achieved, but also to be surpassed. Process enablers and critical success factors must be clearly understood for any permanent improvement to be achieved and sustained. As mentioned in the preface, ISO-55000 considers these enablers and critical success factors as elements. Some of them included:

- Proper resources
- Competent personnel
- Organizational awareness of the buiness objectives
- Good organizational communication
- Information collection, analysis, and utilization
- Good documentation

Properly identifying these enablers and their impacts will require extensive data collection, both internally and from the benchmarking partners.

Enablers

Enablers are a broad set of activities or conditions that help to enhance the implementation of a best business practice. Although benchmarking always compares hard processes, an essential part of a true benchmarking approach also includes analyzing the management skills and attitudes that combine to allow a company to achieve best business practices. This hidden narrative is as important during the benchmarking exercise as are the visible statistical factors.

The enablers, therefore, are behind-the-scene, hidden factors that allow the development or continuation of best practices. Some examples include leadership, motivated workforces, management vision, and organizational focus. These factors have a direct impact on the company's quantified financial performance, but are rarely mentioned by specific statistics. These methods or approaches lead to a company's exceptional performance. Realize that these methods or approaches are relative, not absolute. In other words, they are not perfect; they too can be improved. This highlights a previous note — enablers can be better, but are rarely identified as the best.

Enablers or critical success factors can be found anywhere. They know of no industrial, political, or geographical boundaries.

How do we compare one company to another by enablers? We start with an internal analysis. For companies to be successful, they must have a thorough knowledge and understanding of their internal processes. Without this understanding, it would not be possible to recognize the differences between themselves and their benchmarking partners. It would be virtually impossible to recognize and integrate the differences and innovations that are found in best practice companies.

Defining Core Competencies

Benchmarking, as a continuous improvement tool, is only to be used to improve core business processes. What are core business processes? All the definitions focus on processes that allow companies to differentiate themselves from their competitors. A core business process may have an impact by lowering costs, increasing profits, providing improved service to customers, improving product quality, or improving regulatory compliance.

Several authors have defined core competencies for businesses. In his 1997 text *Operations Management,* Richard Schonberger defines core competencies as "a key business output or process through which an organization distinguishes itself positively." He specifically mentions expert maintenance, low operating costs, and cross-trained labor.

Another author is Gregory Hines who, in his text, The Benchmarking Workbook, defines a core competency as "a key business process that represents core functional efforts and are usually characterized by transactions that directly or indirectly influence the customer's perception of the company". He further lists several processes, including:

- Procuring and supporting capital equpment
- Managing and supporting facilities

As can be seen, the maintenance function clearly Hines's definition of a core business process.

In *The Benchmarking Management Guide,* written by the American Productivity and Quality Center, core competencies are identified as business processes that should impact the following business measures:

- Return on Net Assets
- Customer Satisfaction
- Revenue per employee
- Quality
- Asset Utilization
- Capacity

The maintenance function in any plant or facility fits this definition as well as the previous two.

Still other sources point to a core competency as any aspect of the business operation that results in a strategic market advantage. When evaluating the maintenance process in any company, a core competency point to a strategic market advantage in many ways already mentioned. These include enhancing any quality initiative, increasing capacity, reducing costs, and eliminating waste.

Finally, we have the focus from PAS-55 and ISO-55000. Because maintenance is a key part of the life cycle (Figure P-2), any organization pursuing asset management would have to recognize maintenance as a core business process.

Maintenance and ROFA

The investment a company makes in its assets often is measured against the profits the company generates. This measure is called return on fixed assets (ROFA). Note: Many companies use some form of measure such as ROA (Return on Assets), RONA (Return on Net Assets), etc. These types of indicators are used in strategic planning when a company picks what facility to occupy or the plant in which to produce a product.

Asset management focuses on achieving the lowest total life-cycle cost to produce a product or provide a service. The goal is to have a higher ROFA than your competitor, so as to be the low-cost producer of a product or service. A company in this position attracts customers and ensures greater market share. Furthermore, a higher ROFA will attract investors to a company, ensuring a sound financial base on which to build further business.

It is the responsibility of all departments or functions within a company to measure and control their costs because these costs ulti-

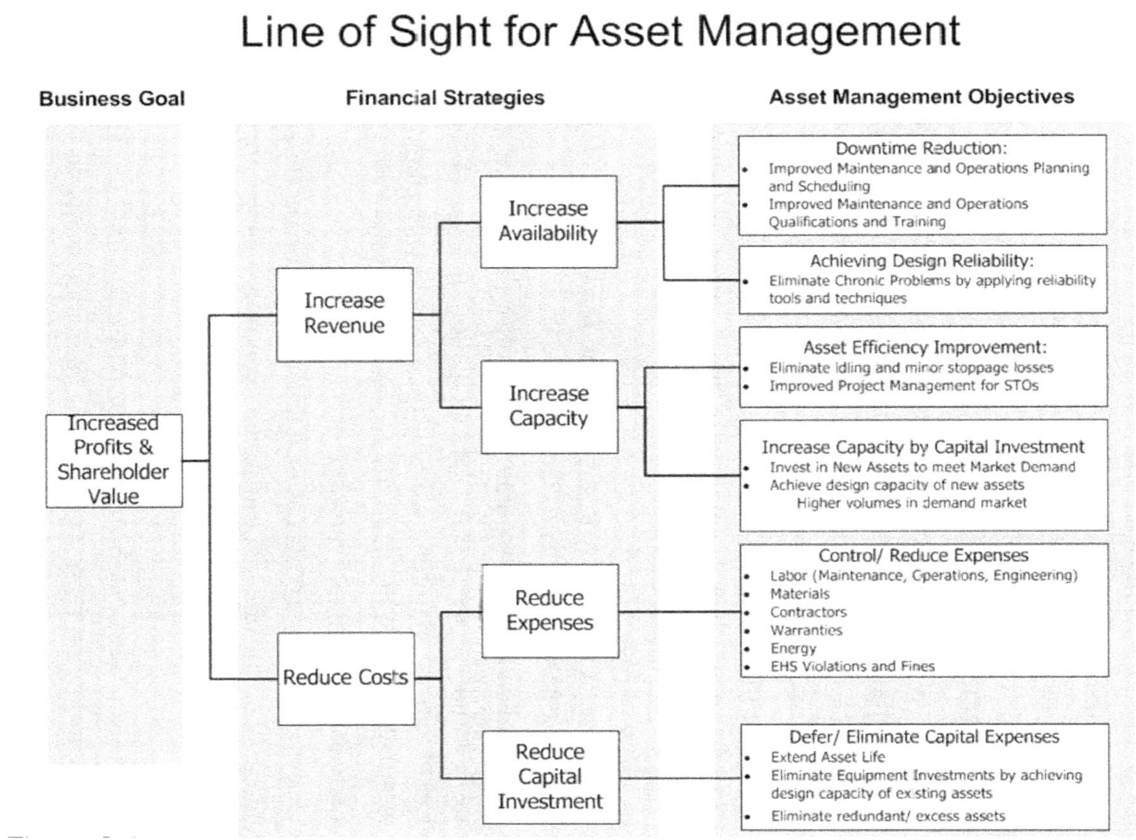

Figure 2-1, Financial Line of Sight

mately will impact the ROFA calculation. It is only when all departments or functions within a company work together that the maximum ROFA is achieved.

If we revisit Figure P-1 (the line of sight diagram), we can reconfigure it to illustrate this point. Figure 2-1 shows the line of sight, with a focus on ROFA.

On the right side of Figure 2-1, the various functional departments that can impact the ROFA calculation are listed. Maintenance, of course, is one of these departments. Because the maintenance function is our focus here, it is beyond the scope of this chapter to deal with all the other areas in detail.

In what ways does maintenance management impact the ROFA calculation? Two indicators may be utilized to show the impact.

Maintenance Costs as a Percentage of Total Process, Production, or Manufacturing Costs

This indicator is an accurate measure for the costs of manufacturing. It should be used as a total calculation, not a per-production-unit calculation. Maintenance will be a percentage of the cost to produce, but is generally fixed. This stability makes the indicator more accurate for the financial measure of maintenance, because it makes trending maintenance costs easier. If the maintenance cost percentage fluctuates, then the efficiency and effectiveness of maintenance should be examined to find the cause of the change.

Maintenance Cost per Square Foot Maintained

This indicator compares the maintenance costs to the total amount of floor space in a facility. This is an accurate measure for facilities because the cost is also usually stable. This indicator is easy to use to trend any increases over time. If the percentage of maintenance costs fluctuates, then the efficiency and effectiveness of maintenance should be examined to find the cause of the change.

These two indicators show that traditional maintenance labor and material costs will have an impact on the ROFA. However, insuring that the equipment or assets are available and operating efficiently can also have an impact on ROFA.

Expenses and Capacity

The total impact of the maintenance function is illustrated in Figure 2-2. Because the maintenance function impacts not only expenses but also the capacity, it has a large role in contributing to ROFA. (Figure 2-3 will further clarify this diagram by detailing the blocks of the pyramid.)

Overall, the goal for any company is to increase profitability. This goal holds whether the company is public and has shareholders, or is private and held by an owner(s). The maintenance or asset management function can increase profits in two main ways: decreasing expenses and increasing capacity.

While considering decreasing expenses, it is estimated that 1/3 of all maintenance expenditures are wasted through inefficient and ineffective utilization of the maintenance resources. There are two main divisions of the maintenance costs: labor and materials. If a maintenance labor budget for a company is \$3M annually, and 1/3 of it is wasted, then \$1M could be saved over time. This savings would not necessarily be in headcount reduction. It may be a reduction in overtime, reduction in the use of outside contractors, or deferral of maintenance without additional expenditures.

If the maintenance labor budget is \$3M annually, then studies show that the materials budget will be a similar amount. If a \$3M materials budget can also be reduced by 1/3, then the savings for improv-

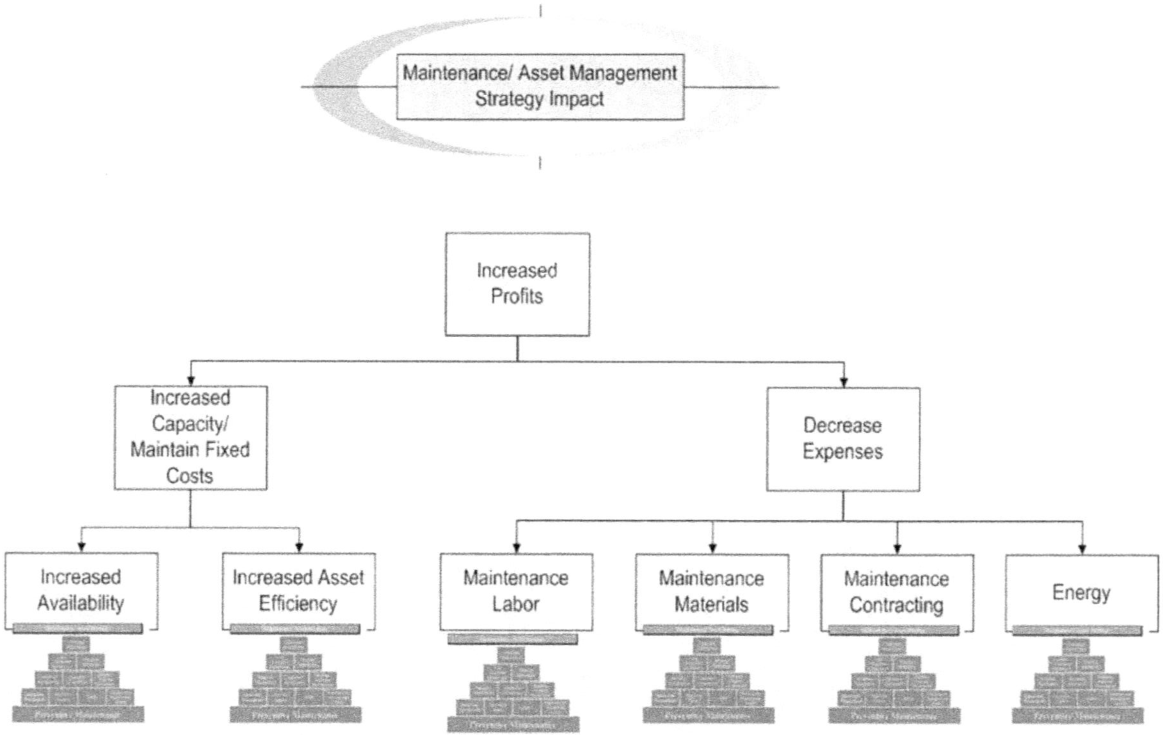

Figure 2-2 Maintenance / Asset Management Strategy Impact

ing maintenance efficiency and effectiveness combined with the materials savings could approach $2M per year. This amount is actually expense dollars that would not be required. Expense dollars that are not used translate to profit dollars.

Realize that, when improving a reactive maintenance organization, these savings are not immediate. It takes some time to realize these total savings because changing a reactive maintenance organization to a proactive, best practice organization can take from 3 to 5 years. It's not that this transition is by itself technically difficult. Instead, time is required to change the corporate culture or paradigm from one of negativity towards the maintenance function to one that truly treats the function as a core business process.

The pure maintenance contribution to profitability is dwarfed when compared to the savings realized by increasing the capacity (availability) and efficiency of the assets being maintained. For example, in some companies, equipment downtime may average 10 to 20%, and even more. Equipment being down, when it is supposed to be operating, restricts the amount of product that is deliverable to the market. Some companies have gone as far as to purchase backup or redundant equipment to compensate for equipment downtime. This has a negative impact on their return on fixed assets indicator and lowers their investment ratings in the financial community.

Even in markets where there is a cap on volume, downtime increases costs and prevents a company from achieving the financial results desired, whether it is to increase profit margins or to be the low cost supplier. There are some organizations that refuse to calculate a cost of downtime and some that even say there is not cost to downtime. However, they fail to consider the following costs, among others:

In reality, the true cost of downtime is the lost sales dollars of the product not made on time. Although this amount is significantly higher than the base costs mentioned previously, it is the true cost of downtime. In most companies, the finance departments end up with a compromise number. However, companies needs to have a clear understanding of this cost to make good decisions concerning their assets and how they are operated.

For example, if the company discussed previously under maintenance savings examined their unplanned downtime for the previous year, they may find that there was a considerable amount, only part of which can be corrected by improving maintenance. Some of the other causes for equipment downtime could be related to raw materials, production scheduling, quality control, operator error, etc. However, if the maintenance downtime alone was valued at $38M (downtime loss costs) and a 50% reduction in these costs was achievable if maintenance was improved, the savings could be $19M. If even $2M of this amount was spent on maintenance to achieve this savings, the gross savings would still be $17M in reduced downtime loss costs (throughput). This line of reasoning relegates the projected maintenance savings previously discussed to a minor issue.

- Utility costs
- Cost of idle production / operations personnel
- Cost of late delivery
- Overtime costs to make up lost production to meet schedules

In addition to the cost of pure downtime is the cost of lost efficiency. One company examined the efficiency of its gas compressors on an off-shore operation. It found the compressors were operating (due to age and internal wear) at 61% efficiency. This loss was approximately $5.4M annually. The overhaul would cost $450K, including labor, materials, and downtime production losses. The decision was made to overhaul the compressors serially, to avoid total shutdown. The compressor overhaul was paid back 28.1 days after restart and the $5.4M in increased production was realized in the next 12 months.

Many Japanese studies (related to total productive maintenance, or TPM) have shown that efficiency losses are always greater than pure downtime losses. However, this fact becomes more alarming when you consider that most efficiency losses are never measured and reported. This leads to many chronic problems that are never solved until a breakdown occurs. Some chronic problems that have a dramatic impact on equipment efficiencies are never discovered. Only when accurate maintenance records are kept are these problems discovered. Then, utilizing the maintenance data combined with the financial data, the root cause of the efficiency problem can be solved.

If asset management is a focus for an organization, it is possible for the maintenance function to contribute to overall plant profitability. Although it takes cooperation and focus of all departments and functions within an organization to be successful, the maintenance department can have a dramatic positive impact on ROFA.

Because maintenance is typically viewed as an expense, any maintenance savings can be viewed as directly contributing to profits. By achieving maximum availability and efficiency from plant assets, a plant or facilities manager insures that a company does not need to invest in excess assets to produce its products or provide its services. The reduction in total assets contributes to overall improvement to ROFA for any company. Additionally, if the true goal of asset management is to maximize the value received from the assets, then a fully functioning maintenance organization is required for PAS-55 or ISO-55000 compliance.

The information in this section clearly establishes maintenance management as a core business process. Therefore, it is a process than could benefit from benchmarking. The next question then is what type of benchmarking should be utilized to gain the maximum benefits?

Types of Benchmarking

There are several types of benchmarking that can be used to conduct a benchmarking project. They are:

1. Internal
2. Similar Industry / Competitive
3. Best Practice

Internal Benchmarking

Internal benchmarking typically considers different departments or processes within a plant. This type of benchmarking has some advantages in that the data can be collected easily. It is also easier to compare the data because many of the hidden factors (enablers) do not have to be closely checked. For example, the departments will have a similar culture, the organizational structure will likely be the same, and the skills of the personnel, labor relations, and management attitude will be similar. This will make data comparison quick and easy.

The disadvantage of this type of benchmarking is that it is unlikely to result in any major breakthrough in improvements. However, internal benchmarking will lead to small, incremental improvements and should provide adequate Return on Investment (ROI) for any improvements that are implemented. The successes in internal benchmarking will very likely result in the desire for more extensive external benchmarking.

Similar Industry / Competitive Benchmarking

Similar industry or competitive benchmarking involves external partners in similar industries or processes. In many benchmarking projects, even competitors are utilized. This may be difficult in some industries, but many companies are open to sharing information, as long as it is not proprietary.

With similar industry / competitive benchmarking, the project tends to focus on organizational measures. In many cases, this type of benchmarking focuses on meeting a "number" rather than improvement of any specific business process. In competitive benchmarking, small or incremental improvements are noted, but paradigms for competitive businesses are similar. This indicates that the improvement process will be slow when relying on competitive or similar industry benchmarking.

Best Practice Benchmarking

Best Practice benchmarking focuses on finding the unarguable leader in the process being benchmarked. This search will cross industry sectors and geographical locations. This approach provides the opportunity for developing "breakthrough strategies" for a particular industry. Organizations study business processes outside their industry, adapt or adopt superior business processes, and make a quantum leap in performance in comparison to their competitors. Being the early adaptor or adopter gives organizations an opportunity to lower costs or aggressively capture market share.

One of the keys to being successful with best practice benchmarking is to define a best practice. Does "best" mean:

- Most efficient?
- Most cost effective?
- Most customer service oriented?
- Most profitable?

When conducting a best practice benchmarking project, an organization must clearly define what is a best practice for that organization. Without this clear understanding, it will take more resources to conduct a benchmarking project; furthermore, the improvements derived from benchmarking will be mediocre at best.

The Best Practices Ad Hoc Committee in the GSA Office of Governmentwide Policy developed the following definition for best practices:

Best Practices are good practices that have worked well elsewhere. They are proven and have produced successful results. They must focus on proven sources of best practices.

It goes on to state:

They should schedule frequent reviews of practices to determine if they are still effective and whether they should continue to be utilized.

This definition indicates that best practices evolve over time. What may have been a best practice in the past is currently a good practice; sometime in the future, it may be a poor practice. Continuous improvement calls for progressive forward movement, not status quo business processes.

When looking for Best Practice companies, understand that no single best practice company will be found. This is simply because all companies have strengths and weaknesses. There are no perfect companies. Because the processes that organizations are trying to improve through benchmarking vary, the companies identified as the Best will also vary. Benchmarking with the best requires systematic and thorough planning and data collection.

When comparing the three type of benchmarking, Best Practice benchmarking is superior. It provides the opportunity to make the most significant improvement because companies being benchmarked are the best in the particular process. Best practice benchmarking provides the greatest opportunity to achieve the maximum return on investment for the benchmarking project. Most important, best practice benchmarking provides the greatest potential for achieving breakthrough strategies, resulting in increasing a company's competitive position.

Benchmarking — The Process

How does the benchmarking process flow? The following are the steps necessary to conduct a successful benchmarking project. They are:

- Conduct Internal Analysis
- Identify Areas for Improvement
- Find Partners
- Make Contact, Develop Questionnaire,and Perform Site Visits
- Compile Results
- Develop and Implement Improvements
- Do It Again...

When conducting an internal analysis, it is important to use a structured format. The analysis may be a survey, such as the one presented in Chapter 1 in this textbook. The goal in conducting the analysis is to identify weaknesses in the organization, areas that need improvement.

Using the analysis in this text, an organization should find the area where it has the greatest deviation from the averages and begin their benchmarking project in that area.

Once the process areas needing improvement are determined, benchmarking partners who are markedly better in the process must be identified. Contacts with these potential partners need to be made to insure the organization is willing to participate in benchmarking.

Next, a questionnaire should be developed, based on the analysis conducted earlier. The questionnaire is sent to the partners; site visits are scheduled and conducted. The information gathered in this process is compiled and analyzed. In turn, recommendations are made for changes to improve the benchmarked process. Once these changes are implemented and improvements noted, the process starts over again.

It is important to conduct an internal analysis before each benchmarking exercise — do not rely on the previous analysis. When one process is improved, it may cascade improvement to other processes. These improvements would not be noted and a process that does not need improvement could be chosen for the next benchmarking project. The benchmarking project would not produce the projected improvements, in which case the organization may no longer view benchmarking as cost effective.

Benchmarking is an evolutionary process. A company may start with internal partners. Incremental improvements will be noted. Based on these results, the progress is made to better practice partners, whether internal or external to the company. With the next round of results in hand and new areas for improvements identified, the next step is benchmarking with the best.

The key to success is always to find a partner who is measurably better in the process you need to benchmark. Once you have achieved process parity with the partner, find a new partner who is still even further along in the process. The benchmarking process continues until the ultimate best is found and superiority over this partner's processes is achieved. This leads to your company's process being identified as the best.

There are NO shortcuts!

Developing a Maintenance Strategy

The focus of the maintenance function is to insure that all company assets meet and continue to meet the design function of the asset. An expansion on the role of the maintenance organization within a company is discussed in Chapter 2.

A definition of best practices adapted to the maintenance process reads: "The maintenance practices that enable a company to achieve a competitive advantage over its competitors in the maintenance process."

What are best practices for the maintenance process in a company today? The practices or processes within maintenance can be discussed in the following 10 categories. They are:

1. Preventive Maintenance
2. Inventory and Procurement
3. Work Flow and Controls
4. Computerized Maintenance Management System Usage
5. Operational Involvement
6. Predictive Maintenance
7. Reliability Centered Maintenance
8. Total Productive Maintenance

9. Financial Optimization
10. Continuous Improvement — Benchmarking

Figure 2-3 illustrates the process relationships.

1. Preventive Maintenance

The preventive maintenance (PM) program is the key to any attempt to improve the maintenance process. This program reduces the amount of reactive maintenance to a level that allows the other practices in the maintenance process to be effective. However, most companies in the United States have problems keeping the PM program focused. In fact, surveys have shown that only 20 percent of U.S. companies believe their PM programs are effective.

Most companies need to focus on the basics of maintenance if they are to achieve any type of best-in-class status. Effective PM activities enable a company to achieve a ratio of 80 percent proactive maintenance to 20 percent (or less) reactive maintenance. Once the ratios are at this level, other practices in the maintenance process become more effective.

2. Inventory and Procurement

The inventory and procurement programs must focus on providing the right parts at the right time. The goal is to have enough spare parts, without having too many spare parts.

No inventory and procurement process can cost-effectively service a reactive maintenance process. However, with the majority of maintenance work planned several weeks in advance, the practices within the inventory and procurement process can be optimized.

Many companies see service levels below 90 percent, which

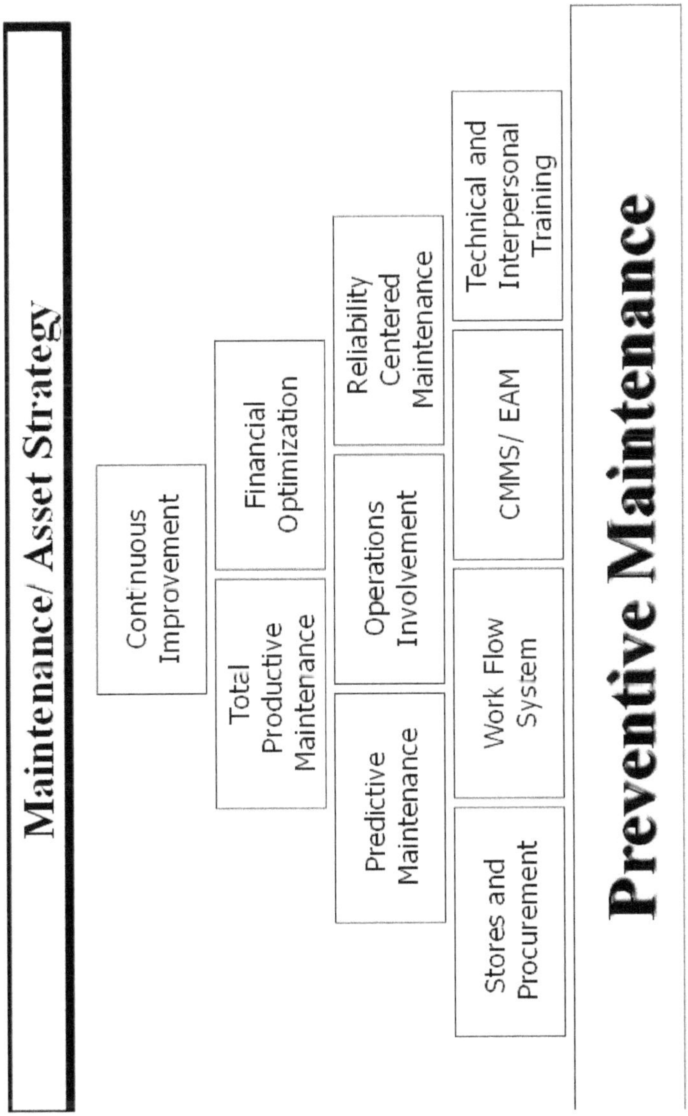

Figure 2-3, The Maintenance Management Pyramid

means stock outs run greater than 10 percent of requests made. This level of service leaves customers (maintenance personnel) fending for themselves, stockpiling personal stores, and circumventing the standard procurement channels to obtain their materials.

To prevent this situation, it is necessary to institute the type of stores controls that will allow the service levels to reach 95 to 97 percent with 100 percent data accuracy. When this level of stores and procurement performance is achieved, you can start the next step toward improvement.

3. Work Flows and Controls

This practice involves documenting and tracking the maintenance work that is performed. It involves the use of a work order system to initiate, track, and record all maintenance activities. The work may start as a request that needs approval. Once approved, the work is planned, then scheduled, performed, and finally recorded. Unless the discipline is in place and enforced to follow this process, data is lost, and true analysis can never be performed.

The solution requires comprehensive use of the work order system to record all maintenance activities. Unless the work is tracked from request through completion, the data is fragmented and useless. If all of the maintenance activities are tracked through the work order system, then effective planning and scheduling can start.

Planning and scheduling requires someone to perform the following activities:

- Review the work submitted
- Approve the work
- Plan the work activities
- Schedule the work activities
- Record the completed work activities

Unless a disciplined process is followed for these steps, productivity decreases and reduced equipment downtime never occurs. At least 80 percent of all maintenance work should be planned on a weekly basis. In addition, the schedule compliance should be at least 90 percent on a weekly basis.

4. Computerized Maintenance Management Systems/ EAM Systems

In most companies, the maintenance function utilizes sufficient data to require the computerization of the data flow. This facilitates the collection, processing, and analysis of the data. The usage of the Computerized Maintenance Management System (CMMS) or Enterprise Asset Management (EAM) System has become popular in most countries around the world. CMMS / EAM software manages the functions discussed previously, and provides support for some of the best practices that are mentioned later in this text .

CMMS / EAM have been used for decades in some countries with very mixed results. Surveys in the United States since 1990 have shown the majority of companies use less than 50 percent of their CMMS/ EAM capabilities. One requirement for a company to be effective in CMMS/EAM usage is complete usage of their system and complete accuracy of the data collected.

5. Operational Involvement

The operations or production departments must take ownership of their equipment to the extent that they are willing to support the maintenance department's efforts. Operational involvement, which varies from company to company, includes some of the following activities:

- Inspecting equipment prior to start up
- Filling out work requests for maintenance
- Completing work orders for maintenance
- Recording breakdown or malfunction data for equipment
- Performing some basic equipment service, such as lubrication
- Performing routine adjustments on equipment
- Executing maintenance activities (supported by central maintenance)

The extent to which operations is involved in maintenance activities may depend on the complexity of the equipment, the skills of the operators, or even union agreements. The goal should always be to free up some maintenance resources to concentrate on more advanced maintenance techniques.

6. Predictive Maintenance

Once the maintenance resources have been freed up because the operations department has become involved, these resources should be refocused on the predictive technologies that apply to their assets. For example, rotating equipment is a natural fit for vibration analysis, electrical equipment for thermography, and so on.

The focus is to investigate and purchase technology that solves or mitigates chronic equipment problems that exist, not to purchase all of the technology available. The predictive maintenance (PDM) inspections should be planned and scheduled utilizing the same techniques that are used to schedule the preventive tasks. All data should be recorded in, or interfaced to, the CMMS.

7. Reliability Centered Maintenance

Reliability Centered Maintenance (RCM) techniques are now applied to the preventive and predictive efforts to optimize the programs. If a particular asset is environmentally sensitive, safety related, or extremely critical to the operation, then the appropriate PM/PDM techniques are decided upon and utilized.

If an asset is going to restrict or impact the production or operational capacity of the company, then another level of PM/PDM activities is applied with a cost ceiling in mind. If the asset was allowed to fail and the cost to replace or rebuild the asset was expensive, then another level of PM/PDM activities would be specified. There is always the possibility that it is more economical to allow some assets to run to failure, and this option is considered in RCM.

The RCM tools require data to be effective. For this reason, the RCM process is used after the organization has progressed to the point that ensures accurate and complete asset data.

8. Total Productive Maintenance

Total Productive Maintenance (TPM) is an operational philosophy whereby everyone in the company understands that their job performance impacts the capacity of the equipment in some way. For example, operations may understand the true capacity of the equipment and not run it beyond design specifications, which could create unnecessary breakdowns.

TPM is like Total Quality Management. The only difference is that companies focus on their assets, not their products. TPM can utilize all of the tools and techniques used to implement, sustain, and improve the total quality effort.

9. Financial Optimization

This statistical technique combines all of the relevant data about an asset, such as downtime cost, maintenance cost, lost efficiency cost, and quality costs. It then balances that data against financially optimized decisions, such as when to take the equipment off line for maintenance, whether to repair or replace an asset, how many critical spare parts to carry, and what the maximum-minimum levels on routine spare parts should be.

Financial optimization requires accurate data, because making these types of decisions incorrectly could have a devastating effect on a company's competitive position. When a company reaches a level of sophistication where this technique can be used, it is approaching best-in-class status.

10. Continuous Improvement

Continuous improvement is best epitomized by the expression, "best is the enemy of better." Continuous improvement in asset care is an ongoing evaluation program. This includes constantly looking for the "little things" that can make a company more competitive.

Benchmarking is one of the key tools for continuous improvement. Of the several types of benchmarking practices, the most successful is Best Practice benchmarking, which examines specific processes (most likely the 10 just mentioned) in maintenance, compares the processes to companies that have mastered those processes, and maps changes to improve the specific process. This flow of practices in maintenance is important, because understanding that Benchmarking is a Continuous Improvement tool enhances the understanding that it is a technique employed by a mature organization, one that is knowledgeable about the maintenance business process.

Key Performance Indicators (KPIs), Benchmarking, and Best Practices

Performance indicators, or measures, for best practices are misunderstood and misused in most companies. Properly used, performance indicators should highlight opportunities for improvement within companies today. Performance indicators should highlight a "soft spot" in a company, then enable further analysis to find the problem that is causing the low indicator, and then ultimately point to a solution to the problem.

Performance indicators are valuable tools in highlighting areas that are potential processes to be benchmarked. For example, if a certain set of performance indicators show that a maintenance process, such as preventive maintenance needs to be improved, and the internal personnel for the company can not identify the changes necessary to improve, then a benchmarking project may be the answer.

However, it is necessary to clarify that benchmarks are not performance indicators and performance indicators are not benchmarks. Using performance indicators is an internal function for a company. A benchmark is an external goal that is recognized as an industry or process standard. However, the number in itself is meaningless, unless there is an understanding of how the benchmark is derived. Understanding the enablers and success factors behind the benchmark is what is important.

Also, it must be clearly understood that there is a difference between a benchmark and the process of benchmarking. The benchmark is again a number. Benchmarking is a process of understanding a company's processes and practices, so they can be adapted or modified, and then adopted by a company, in order to be superior in the process or practice being studied.

CONTINUOUS IMPROVEMENT — THE KEY TO COMPETITIVENESS

Because benchmarking is a continuous improvement tool, it should only be started if a company wants to make changes to improve. Companies can not develop the attitude "We have always done it this way." They must be willing to change to meet the challenges of increasing competitive pressure.

Benchmarking is a continuous improvement tool that can facilitate change. As Best Practice companies are examined and their processes understood, the gap between a company's present practices and Best Practice promotes dissatisfaction and desire for change. Seeing, understanding, and learning from Best Practice companies helps companies identify what to change and how to make the changes to maximize their return on their investment in the changes. Witnessing Best Practices provides a realistic and achievable picture of the desired future. However, this takes resources, both in human and financial capital to be successful. It is necessary to explore with the benchmarking partners, the tangible and intangible factors that combine to produce superior performance. It is also necessary to involve those people most directly connected with the business process being benchmarked, because they have to take ownership in the changed process.

Conducting a Benchmarking Project

In considering how to conduct a benchmarking project, it is necessary to review the goals of benchmarking. Benchmarking should:

1. Provide a measure for the benchmarked process
    a. This measure allows for an "Apples to Apples" comparison
2. Clearly describe the organization's performance gap when compared to the measure

3. Clearly identify the Best Practices and enablers that produced the superior performance observed during the benchmarking project
4. Set performance improvement goals for the bench marked processes and identify actions that must be taken to improve the process

Quantifying the organization's current performance, the Best Practice for the process, and the performance gap is vitally important. There is a management axiom that says:

"If you don't measure it, you don't manage it."

This is true of benchmarking. There must be quantifiable measures if a clear strategy to improve is going to be developed. This details the "SMART" requirements for a benchmarking project. The acronym SMART means:

1. Specific — this insures the project is focused.
2. Measurable— this requires quantifiable measures.
3. Achievable — This insures that it is within a business objective.
4. Realistic — Again, focused on a business objective.
5. Time framed — The benchmarking project should have a start and end date.

Gap Analysis

One of the key components of a benchmarking project that helps the project achieve the SMART objectives is a Gap Analysis. The Gap Analysis is divided into three main phases. They are:

1. Baseline —the foundation or where the company is at the present time.
2. Entitlement —the best that the company can achieve with effective utilization of their current resources
3. Benchmark —the Best Practice performance of a truly optimized process

To be effective in utilizing the Gap Analysis, the benchmarking project must be able to produce quantifiable results. All of the measures must be able to be expressed clearly and concisely so that the improvement program can be quantified.

The first step of the Gap Analysis is to compare the company's process in quantifiable terms to the Best Practice results that were observed. It is best to plot this information as shown in Figure 2-4.

Figure 2-4, Gap Analysis

Note that the gap between the observed Best Practice and the organization's current performance is plotted on the vertical axis of the chart. The horizontal axis shows the time line. Figure 2-4 highlights the need for the measures to be quantifiable if they are going to be properly graphed on the Gap Analysis.

The second part of the Gap Analysis sets the time (T1) to reach what is called a current parity goal. This goal focuses on achieving the current level of performance that the Best Practice Company has reached at the current time. This graph also recognizes that the Best Practice Company will have made improvements during this time period and will still be at a higher level of performance. This relationship is shown in Figure 2-5.

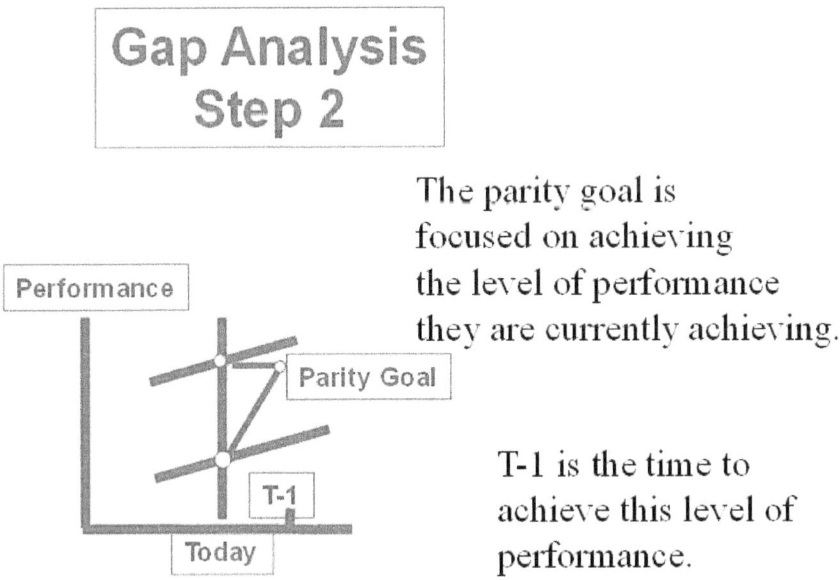

Figure 2-5, Gap Analysis Step 2

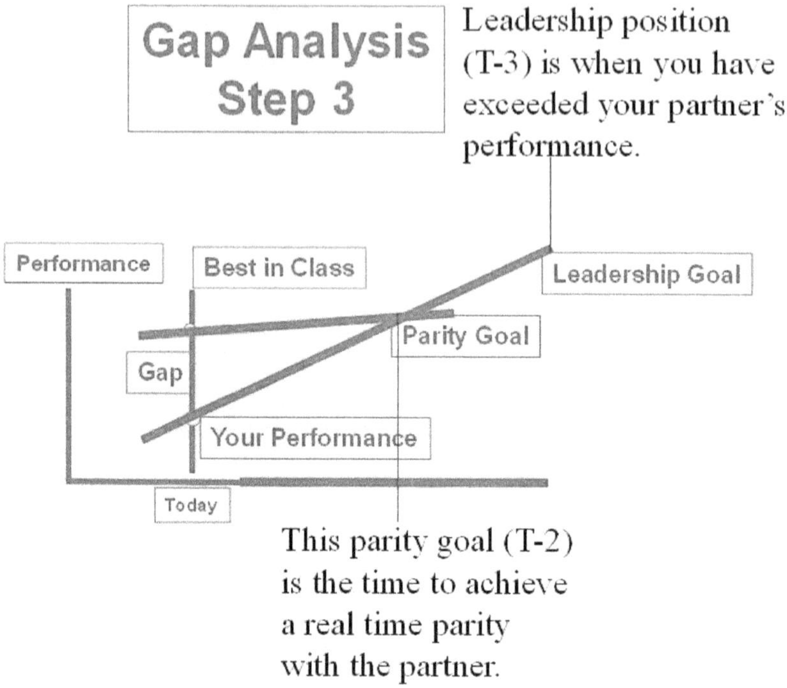

Figure 2-6, Gap Analysis Step 3

The next step is to set a real time parity goal. During this step, your company will achieve parity on the benchmarked process with the Best Practice Company. This time is highlighted in Figure 2-6 as T-2. The final goal is the leadership position, which occurs when your company's performance in the benchmarked process is recognized as having exceeded your partner's performance. At this time, noted on the graph as T-3, your company will be recognized as the Best Practice company when it comes to the benchmarked process.

If a company is to be effective in utilizing the Gap Analysis, all of the parameters must be quantifiable and time framed. Otherwise, the Gap Analysis would be meaningless.

The Benchmarking Process

What process should be used to insure that a benchmarking process is quantifiable? The following checklist can be used. If it is followed, it allows for the benchmarking process to be successful. If a disciplined approach is not utilized, it is unlikely that the benchmarking process will produce any long-term results. The Benchmarking Checklist is as follows:

1. Plan
2. Search
3. Observe
4. Analyze
5. Adapt
6. Improve

The checklist can then be further expanded into the following detail.

1. PLAN

A. What are our Maintenance Mission or Goals and Objectives?

i. This information insures that everyone involve has a clear understanding of the maintenance business function.

B. What is our maintenance process?

i. This step involve work flows, business process flows, etc.

C. How is our maintenance process measured?

i. What are the current KPIs or performance indicators?

D. How is our maintenance process perceived as performing today?

i. What is the level of satisfaction for the service that maintenance performs?

E. **Who is the perceived customer for maintenance?**
   i. Is it operations or the shareholders/ owners? The answer to this question can prove insightful in determining the level of understanding of maintenance within the organization and the time it will take to make improvements.

F. **What services are expected from the maintenance function?**
   i. What service does maintenance perform? What is outside contracted? What isn't being done that needs to be done?

G. **What services is the maintenance function prepared to deliver?**
   i. Are they capable of more? Is the staffing, skill levels, etc. at the correct level to perform the services?

H. **What are the performance measures for the maintenance function?**
   i. How does maintenance know if they are achieving their objectives?

I. **How were these measures established?**
   i. Were they negotiated or mandated?

K. **What is the perception of our maintenance function compared to our competitors?**
   i. Internal perceptions — worse than, as good as, or better than?

2. SEARCH

A. **What companies are better at a maintenance process than our company?**
   i. Utilize magazine articles and internet sites

B. **Which company(s) is considered to be the best?**
   i. Consider NAME (North American Maintenance Excellence) Award Winners—http://www.nameaward.com

C. **What can we learn if we benchmark with this company?**
   i. Understand their Best Practices and how they could help your company

D. **Who should we contact to determine if they are a potential benchmarking partner?**
   i. Look for a contact in the article or on the internet site

3. OBSERVE

A. **What are their Maintenance Mission or Goals and Objectives?**
   i. How does this compare to your company's?

B. **What are their performance measures?**
   i. How do they compare to your company's?

C. **How well does their maintenance strategy perform over time and/ or at multiple locations**
   i. Are their current results an anomaly or are they sustainable?

D. **How do they measure their maintenance performance?**
   i. Are their measures different from your company's?

E. **What enables their Best Practice performance in maintenance?**
   i. Plant Manager, Corporate Culture, etc?

F. **What factors could prevent our company from adopting their maintenance policies and practices into our maintenance organization?**
   i. Your culture, work rules, maintenance paradigm, etc?

4. ANALYZE

A. **What is the nature of the performance gap?**
   i. Comparing their Best Practice to your practice

B. **What is the magnitude of the performance gap?**
   i. How large is the benchmark gap?
C. **What characteristics distinguish their processes as superior?**
   i. Detail the enablers you discovered
D. **What activities do we need to change to achieve parity with their performance?**
   i. This is the plan for change

5. ADAPT

A. **How does the knowledge we gained about their maintenance process enable us to make changes to improve our maintenance process?**
   i. What do you need to do to improve?
B. **Should we adjust, redefine, or completely change our performance measures based on the Best Practices that were observed?**
   i. What were the differences and how could you benefit by the change?
C. **What parts of their Best Practice maintenance processes would have to be changed or modified to be adapted into our maintenance process?**
   i. You need to be an adaptor – not a copy cat

6. IMPROVE

A. **What have we learned that would allow our company to achieve superiority in the maintenance process that was benchmarked?**
   i. What can you change to eventually achieve the superiority position?

B. How can these changes be implemented into our maintenance process?

i. Develop the implementation plan

C. How long should it take for our company to implement these changes?

i. Timeline the implementation plan

To gain maximum benefits from benchmarking, a company should conduct a benchmarking exercise only after it has attained some level of maturity in the core competency being benchmarked. Clearly, a company would have to have some data about its own process before it could perform a meaningful comparison with another company. For example, in equipment maintenance management, common benchmarks are:

1. Percent of maintenance labor costs spent on reactive activities versus planned and scheduled activities.
2. Service level of the storeroom — percent of time the parts are in the storeroom when needed.
3. Percentage of maintenance work completed as planned.
4. Maintenance cost as a percentage of the estimated replacement value of the plant or facility equipment.
5. Maintenance costs as a percentage of sales costs.

Without accurate and timely data and an understanding of how the data is used to compile the benchmark statistics, there will be little understanding of what is required to improve the maintenance process. And this is true whatever process is benchmarked.

When partnering with companies considered to be the best in a certain aspect of a competency, it is also important to have an example of an internal best practice to share with them. Benchmarking requires a true partnership, which includes mutual benefits. If you are only looking and asking during benchmarking visits—with no sharing—what is the benefit to the partners?

The final step to ensure benefits from benchmarking is to use the knowledge gained to make changes in the competency benchmarked. The knowledge gained should be detailed enough to develop a cost/benefit analysis for any recommended changes.

Benchmarking is an investment. The investment includes the time and money to complete the ten steps described earlier. The increased revenue generated by the implemented improvements pays for the investment. For example, in equipment maintenance, the revenue may be produced through increased capacity (less downtime, higher throughput) or reduced expenses (efficiency improvements).

The revenue is plotted against the investment in the improvements to calculate the return on investment (ROI). To ensure success, the ROI should be calculated for each benchmarking exercise.

Benchmarking Code of Conduct

1. Keep it legal.
2. Be willing to give any information you get.
3. Respect confidentiality.
4. Keep the information internal.
5. Use benchmarking contacts.
6. Don't refer without permission.
7. Be prepared from the start.
8. Understand your expectations.
9. Act in accordance with your expectations.
10. Be honest.
11. Follow through with commitments.

Although this list of suggestions for the code of conduct may seem to be common sense, it is surprising the number of companies that fail to apply them. This results in everything from minor disagreements between individuals to major legal battles. Recognizing that the other companies are your partners and treating them as such is key to successful benchmarking relationships.

Traps to Benchmarking

When benchmarking is used properly, it can make a major contribution to the continuous improvement process. However, it can also be completely devastating to a company's competitive position when used improperly. Some of the improper uses of benchmarking follow:

1. Using benchmarking data as a performance goal.

When companies benchmark their core competencies, they can easily fall into the trap of thinking a benchmark should be a performance indicator. For example, they focus all of their efforts on cutting costs to reach a certain financial indicator, losing focus on the real goal.

A company receives greater benefits when the tools and techniques used by a partner to achieve a level of performance are understood. This allows the company not only to meet a certain number, but also to develop a vision of how to achieve an even more advanced goal.

By focusing on reaching a certain number, some companies have changed their organizations negatively (by downsizing or cutting expenses). However, they have also removed the infrastructure (people or information systems) and soon find they are not able to sustain or improve the benchmark. In such cases, benchmarking becomes a curse.

2. Premature benchmarking.

When a company attempts to benchmark before the organization is ready, it may not have the data to compare with its partners. So, someone makes a "guesstimate" that does the company no good.

The process of collecting data gives an organization an understanding of its core competencies and how it currently functions. Premature benchmarking will lead back to the first trap — just wanting to reach a number. Companies stepping into this trap become "industrial tourists." They go to plants and see interesting things, but don't have enough of an understanding to apply what they see to their own businesses. Then, the end results are reports that sit on shelves and never contribute to improved business processes.

3. Copycat benchmarking.

Imitation benchmarking occurs when a company visits its partners and, rather than learning how the partners changed their businesses, concentrates on how to copy the partners' current activities. This is detrimental to a company because it may not have the same business drivers as its benchmarking partners. Also, there may be major constraints to implementing the partner's processes. Such constraints might include incompatible operations (7 days @ 24 hr/day versus 5 days @ 12 hr/day), different skill levels of the work force, differences in union agreements, different organizational structures, and different market conditions.

4. Unethical benchmarking.

Sometimes a company will agree to benchmark with a competitor and then try to uncover proprietary information while on the site visit or by use of the questionnaire. Clearly, this kind of behavior will lead to problems between the companies and virtually ruin any chance of conducting a successful benchmarking exercise at a later date.

A second type of unethical benchmarking involves using the benchmarking partners' names or data in public without receiving prior permission. This, too, will damage any chance for ongoing benchmarking between the companies. Even worse, the bad experience may prevent management from ever commissioning further benchmarking exercises with other partners.

Other Pitfalls

Not every company is ready for benchmarking. However, it is a trap not to do it because of a previous bad experience or because of an attitude of "We are already the best" or "We are different than everyone else." Companies in which responsible individuals have such a mindset will have little chance of improving.

Benchmarking is valuable because trying to reinvent the wheel is an expensive way to try to make improvements. Once a company has the proper view of the benchmarking process, and disciplined guidelines are established and followed, desired improvements should follow. However, if the company does not benchmark for the right reasons, benchmarking efforts will become a curse.

Procedural Review

Benchmarking opportunities are uncovered when a company conducts an analysis of its current policies and practices. Benefits are gained by following a disciplined process, composed of 10 steps:

1. **Conduct an internal audit of a process or processes.**
   a. Education of key personnel in Benchmarking processes is crucial at this point. They must fully understand and support the process.

2. **Highlight potential areas for improvement.**
   a. This requires understanding the cost of Benchmarking compared to the financial benefits that will be derived. This should be presented in a return on investment business case.
3. **Do research to find 3 or 4 companies with superior processes in the areas identified for improvement.**
4. **Contact those companies and obtain their cooperation for benchmarking.**
5. **Develop a pre-visit questionnaire highlighting the identified areas for improvement. (See Step 2.)**
   a. This step requires a carefully planned approach to Bench marking. You then will need the discipline to adhere to the plan.
6. **Perform the site visits to the 3 or 4 partners. (See Step 3.)**
   a. An interim report should be prepared after each visit and presented to the executive sponsor.
7. **Perform a gap analysis on the data gathered compared to your company's current performance. (See Figure 2-4.)**
8. **Develop a plan for implementing the improvements.**
   a. This includes the changes required, personnel involved, and the timeline
9. **Facilitate the improvement plan.**
   a. This means that one or more members of the benchmarking project team will oversee the implementation of the plan to insure the changes are properly implemented.
10. **Start the benchmarking process over again (i.e., go back to Step 1).**

Benchmarking helps companies find the opportunities for improvement that will give them a competitive advantage in their marketplaces. However, the real benefits from benchmarking do not occur until the findings from the benchmarking project are implemented and improvements are realized.

Final Points

1. It is necessary to explore the tangible and intangible factors that combine to produce a superior performance and involve those people most directly concerned in the activity being examined.

2. Benchmarks are not the end-all. A benchmark performance does not remain a standard for long. Continuous improvement must be the goal.

After having examined the benchmarking process, it is necessary to clearly understand the process being benchmarked. Chapters 3 though 11 will examine all aspects of the maintenance management function. These chapters will further highlight the methodology behind the analysis that was included in Chapter 1. Chapter 12 will then present some current industry benchmarks for maintenance. With the understanding of the benchmarking process and the maintenance process, any company should be able to conduct a successful benchmarking project.

CHAPTER 3

Maintenance Organizations

The goals and objectives of the maintenance organization determine the type of maintenance organization that is established. If the goals and objectives are progressive and the maintenance organization is recognized as a contributor to the corporate bottom line, variations on some of the more conventional organizational structures can be used.

Goals and Objectives of Maintenance Organizations

The typical goals and objectives for a maintenance organization are summarized in Figure 3-1 on the following page.

MAXIMIZE PRODUCTION

Maximize production at the lowest cost, the highest quality, and within the optimum safety standards. This statement is very broad, yet maintenance must have a proactive vision to help focus its activities. The statement should be tied to any corporate objectives. This is especially true if a company is working towards an asset management certification, such as ISO-55000. If this is the case, it should likely

FIGURE 3-1 **OBJECTIVES OF MAINTENANCE MANAGEMENT**
▶ Maximum production at the lowest cost, the highest quality, and within the optimum safety standards. ♦ Delivery of maximum value from the assets.
▶ Identify and implement cost reductions.
▶ Provide accurate equipment maintenance records.
▶ Collect necessary maintenance cost information.
▶ Optimize maintenance resources.
▶ Optimize capital equipment life.
▶ Minimize energy usage.
▶ Minimize inventory on hand.

focus on maximizing value from the assets. In either case, it can be broken down into smaller components.

a. Maintaining existing equipment and facilities

This activity is the primary reason for the existence of the maintenance organization. The organization gains no advantage from owning equipment or facilities that are neither operational nor functional. This component is the "keep-it-running" charter of maintenance.

b. Equipment and facilities inspections and services

These programs are generally referred to as preventive and predictive maintenance. This activity increases the availability of the equipment and facilities by reducing the number of unexpected breakdown or service interruptions.

c. Equipment installations or alterations

Generally, installing and altering equipment are not the charge of the maintenance organization; they are usually performed by outside contract personnel. However, maintenance must still maintain the equipment. Therefore, they should be involved in any equipment installations or alterations.

Identify and Implement Cost Reductions

Reducing costs is sometimes an overlooked aspect of maintenance. However, a maintenance organization can help a company reduce costs in many ways. For example, a change in a maintenance policy may lengthen production run times without damaging the equipment. This change reduces maintenance cost and, at the same time, increases production capacity. By examining its practices, maintenance can usually make adjustments in tools, training, repair procedures, and work planning, all of which can reduce the amount of labor or materials that may be required to perform a specific job. Any time gained while making repairs translates into reduced downtime or increased availability. Downtime is more costly than maintenance expenditures. Before making adjustments made to reduce costs, studies should be conducted to show the before-and-after results. This quantifying of improvements builds management support for maintenance activities.

Provide Accurate Equipment Maintenance Records

Providing accurate equipment maintenance records enables a company to accurately track equipment in such engineering terms as mean time between failure (MTBF) or mean time to repair (MTTR). Success in this endeavor, however, requires accurate records of each maintenance repair, the duration of the repair, and the run-time between repairs. Larger organizations, for which this activity produces a tremendous amount of paperwork, typically use some form of a computerized maintenance management system (CMMS) to track this information. But whether or not a computer is used, all of the maintenance data must be accurately tracked.

This objective seems almost impossible to achieve at times. Maintenance records are generally collected as work orders; they then must be compiled into reports showing meaningful information or trends. The problem is finding enough time to put valuable information on each individual work order. Because excessive amounts of maintenance are performed in a reactive mode, it is difficult to record events after the fact. For example, recording how many times a circuit breaker for a drive motor was reset in one week might seem somewhat insignificant to record on a work order. But, if the overload was due to an increased load on the motor by a bearing wearing inside the drive, it could be analyzed and repaired before the equipment experienced a catastrophic failure. Accurate recordkeeping is mandatory if maintenance is going to fully meet its responsibilities.

Collect necessary maintenance cost information

Collecting necessary maintenance-cost information enables companies to track engineering information. For example, by using life-cycle costing information, companies can purchase assets with the lowest life-cycle costs rather than lowest initial costs. In order to track overall life-cycle costs accurately, all labor, material, contracting, and other miscellaneous costs must be tracked accurately at the equipment level. This tracking is primarily an activity for the maintenance department. However, if an organization is setting ISO-55000 as a goal, this data will need to be accessible to all departments attempting to maximize the value derived from the assets..

In addition to life cycle costing is the need for maintenance budgeting. If accurate cost histories are not collected, how can managers budget what next year's expenses will or should be? Maintenance managers cannot simply say to senior management, "We want to reduce maintenance labor by 10% next year," when they don't really know how the labor resources were allocated this year. Also, if labor figures are only available in dollar amounts, the differences in pay scales may make it difficult to determine how much labor was used in total hours by craft. The information must be collected both in dollars and in hours by craft.

Where is this information collected? Collecting the cost information is again tied to work order control. Knowing the hours spent on the work order multiplied by the labor rates of the individuals performing the work allows a more accurate calculation of the total labor costs used for the work order. Adding up these charges over a given time period for all work orders provides the total labor used. Adding up the hours spent by each craft provides an even clearer picture of the labor resources needed. Material costs can also be determined by tracking what parts were used on the job to each work order. Multiplying the number of parts times their dollar value (obtained from stores or purchasing) calculates the total material dollars spent for a given time period. Contractor and other cost information also must be collected at a work order level.

Each work order form should have the necessary blanks for filling in this information. Only by tracking the information at the work order level can you roll up costs from equipment to line to department to area and, finally, to total plant. Collecting the information at this level also provides cost information for equipment types, maintenance crafts, and cost centers. By utilizing the data gathered through the work order, detailed maintenance analysis reports can be produced, as will be discussed in Chapter 9.

Optimize Maintenance Resources

Optimizing maintenance resources includes eliminating waste through effective planning and scheduling techniques. In reactive maintenance organizations, up to one-third of maintenance expenditures are often wasted. By optimizing maintenance resources, organizations improve their effectiveness in eliminating this waste. For example, if an organization has a maintenance budget of one million dollars and operates in a reactive mode, it is possible that the organization is wasting $300,000. When 80 to 90% of all maintenance activities are planned and scheduled on a weekly basis, there is very little waste to the maintenance process. The goal for a reactive organization is to achieve this level of proficiency.

Optimizing maintenance resources also has an effect on maintenance manpower. For example, with good planning and scheduling practices, a reactive maintenance organization may increase the "wrench time" of their craft technicians from 25% to as much as 60%. This reduces the amount of overtime or outside contracting that an organization currently utilizes, reducing the overall maintenance cost. These types of reductions, while improving service, are essential to optimizing the present resources. Optimizing maintenance resources can only be achieved by good planning and scheduling practices. The disciplines necessary to develop these controls will be discussed in Chapter 6.

Optimize Capital Equipment Life

Optimizing the life of the capital equipment means maintaining the equipment so that it lasts 30 to 40% longer than poorly maintained equipment. The maintenance department's goal is to keep the equipment properly maintained to achieve the longest life cycle. In this way, they are maximizing the value derived from the equipment (asset). A preventive maintenance program designed for the life of the equipment is key to obtaining a maximum life cycle. The maintenance department will then need to perform the correct level of preventive maintenance, performing enough maintenance, but without performing excessive maintenance.

One way to determine a problem in this area is to examine new equipment purchases. Are equipment purchases used to replace equipment in kind? If so, could the purchase of the equipment have been deferred if proper maintenance had been performed on the older equipment? If long life cycles are not being achieved, then the proper level of maintenance is not being performed, and maintenance tasks should be revised. How to achieve this goal is covered in Chapter 7.

Minimize Energy Usage

Minimizing energy usage is a natural result of well-maintained equipment, which requires 6 to 11% less energy to operate than poorly-maintained equipment. These percentages, established by international studies, indicate that maintenance organizations would benefit from constantly monitoring the energy consumption in a plant. Most plants and facilities have equipment that consumes considerable energy if not properly maintained. For example, heat exchangers and coolers that are not cleaned at the proper frequency consume more energy. HVAC systems that are not properly maintained require more energy to provide proper ventilation to a plant or facility.

Even small things can have a dramatic impact on energy consumption. For example, equipment with a poor maintenance schedule will have bearings without proper lubrication or adjustment, couplings not properly aligned, or gears misaligned, all of which contribute to poor performance and require more energy to operate. The key to achieving this objective is having good preventive and predictive maintenance schedules. Setting up an effective PM program is covered in Chapter 7.

MINIMIZE INVENTORY ON HAND

Minimizing inventory on hand helps maintenance organizations eliminate waste. Approximately 50% of a maintenance budget is spent on spare parts and material consumption. In organizations that are reactive, up to 20% of spare parts cost may be waste. As organizations become more planned and controlled, this waste is eliminated. Typical areas of waste in the inventory and purchasing function include:

1. Stocking too many spare parts
2. Expediting spare part delivery
3. Allowing shelf life to expire
4. Single line item purchase orders
5. Vanished spare parts

It is important for the maintenance organization to focus on controlling spare parts and their costs.

Although the goals discussed thus far do not form a comprehensive, all-inclusive list, they highlight the impact a proactive maintenance organization can have on a company. Maintenance is more than a "fix it when it breaks" function. Unless the maintenance organization works with a proactive list of goals and objectives, it will always be sub-optimized.

Management and Maintenance

In the past thirty years, executive management has focused increasingly on short-term profitability, sacrificing their physical assets to do so. Best Practice companies have taken advantage of this trend to develop strategic plans, building strong, complete organizations. One of the foremost areas of focus for these companies has been the maintenance/asset management function. Maintenance is extremely important to being competitive in the world market. But have the majority companies followed their lead? The answer for the majority, sadly, is no. I have seen plants where one day the maintenance force is required to work on sophisticated electronic systems and the next day to perform janitorial service in the lavatories.

In this environment, it is difficult for maintenance personnel to develop a positive attitude of their value to the corporation. If the maintenance function is to become a contributing factor to the survival of companies, management must change their views toward maintenance. If they do, they can achieve world class competitiveness. Achieving the goals necessary to have a strong maintenance organization—one that contributes to increased profitability—will require decisions concerning the maintenance organization and the type of service it provides.

Equipment Service Level

Equipment service level indicates the amount of time the equipment is available for its intended service. The amount of service re-

Figure 3-2 **Maintenance Philosophies**
▶ Reactive Maintenance
▶ Corrective Maintenance
▶ Preventive Maintenance
▶ Predictive Maintenance
▶ Maintenance Prevention

quired from the equipment, along with its resultant costs, determines the type of maintenance philosophy a company will adopt. Five common philosophies are listed in Figure 3-2.

Reactive Maintenance

In far too many cases, equipment is run until it breaks down. There is no preventive maintenance; the technicians react, working only on equipment that is malfunctioning. This approach is the most expensive way to coordinate maintenance. Equipment service level is generally below acceptable levels, and product quality is usually impacted.

Corrective Maintenance

Corrective maintenance activities are generated from PM inspections, routine operational requests, and routine service requirements. These activities make up the maintenance backlog and should be planned and scheduled in advance. This approach is the most cost effective way to perform maintenance, reducing performance costs by 2 to 4 times compared to reactive maintenance. When the majority of maintenance activities fall into this category, equipment service levels can be maintained.

Preventive Maintenance

Preventive maintenance includes the lubrication program, routine inspections, and adjustments. Many potential problems can thus be corrected before they occur. The methods for organizing and developing PM programs will be detailed in Chapter 7. At this level of maintenance, equipment service levels enter the acceptable range for most operations.

Predictive Maintenance

Predictive maintenance allows failures to be forecast through analysis of the equipment's condition. The analysis is generally conducted through some form of trending of a parameter, such as vibration, temperature, and flow. Preventive maintenance differs from predictive maintenance in that it focuses on manual tasks whereas predictive maintenance uses some form of technology. Predictive maintenance allows equipment to be repaired at times that do not interfere with production schedules, thereby removing one of the largest factors from downtime cost. The equipment service level will be very high under predictive maintenance.

An extension of predictive maintenance is condition-based maintenance, which is maintenance performed as it is needed, with the equipment monitored continually. Some plants have the production automation system directly connected to a computer system in order to monitor the equipment condition in a real-time mode. Any deviation from the standard normal range of tolerances will cause an alarm (or in some cases a repair order) to be generated automatically. This real-time trending allows for the maintenance to be performed in the most cost effective manner. Condition-based maintenance is the optimum maintenance cost vs. equipment service level method available. The startup and installation cost can be very high. Nevertheless, many companies are moving toward this type of maintenance.

Maintenance Prevention

Maintenance prevention activities focus on changing the design of equipment components so they require less maintenance. This type of maintenance uses the data gathered from the previous techniques to change the design so the equipment requires less maintenance to perform its design functions for its life. An analogy of an automobile can be used. If the current day auto is compared to a 1970's vintage auto, a reduction in the maintenance requirements can be clearly seen. Tune ups are one of the main areas. 1970's autos required tune ups every 30 to 40 thousand miles. New models require tune ups at 100 thousand miles, with no degradation in performance. These improvements were studied, reengineered, and implemented. Plant and facility equipment today are no different. Maintenance prevention activities usually are supported by the maintenance engineering group.

MAINTENANCE STAFFING OPTIONS

Staffing is an important component of any maintenance organization. Four methods are commonly used to staff the maintenance organization (see Figure 3-3).

FIGURE 3-3 MAINTENANCE STAFFING OPTIONS
▶ Complete In-House Staff
▶ Combined In-House / Contract Staff
▶ Contract Maintenance Staff
▶ Complete Contract Maintenance

COMPLETE IN-HOUSE STAFF

Having a complete in-house staff is the traditional approach in most U.S. companies. Under this approach, the craft technicians who perform maintenance are direct employees of the company. All administrative functions for each employee, as well as salary and benefits, are the responsibility of the company.

Combined In-House/Contract Staff

Combined in-house/contract staff became a more common approach to maintenance in the 1980s. The in-house staff performs most of the maintenance, but contractors perform certain maintenance tasks such as service on air conditioners, equipment rebuilds, and insulation. This method can reduce the amount of staff required for specific skill functions. If the contract personnel are not required full time, this approach can contribute even further savings.

Contract Maintenance Staff

Contract maintenance staffs combine the company's supervisors with contract employees. This method, common in Japan, is gaining popularity in the United States. The contractor provides properly-skilled individuals, removing the burden of training and personnel administration from the company. The downside of the approach is not having the same employees all of the time. Contract employees may have less familiarity with the equipment, but the interaction between the in-house supervision and the contract personnel can help to compensate.

Complete Contract Maintenance

Complete contracting maintenance staff includes all craftsmen, planners, and supervisors. Supervisors generally report to a plant engineer or plant manager. This approach eliminates the need for any in-house maintenance personnel. Although not yet popular in the United States, coupled with an operator-based PM program (explained in the PM section), this program can prove to be cost effective and a valid alternative to conventional maintenance organizations.

In reality, any of the above options can work. In most companies, however, it is difficult to manage a contract work force. Although some companies claim financial benefits from contracting out all maintenance activities, those benefits are imaginary. The perceived benefits occur because the contractor can manage its maintenance work force, whereas the company cannot manage its own. When companies claim a large savings from contracting maintenance, it is typically because they were not efficient and effective in the way they managed their maintenance. After all, the same work gets done. But how can a contractor be cheaper than in-house? Only by planning, scheduling, and removing waste from the maintenance process can the contractor be more cost-effective. Could not the company then, with an internal or in-house work force, achieve the same cost levels?

Another problem comes to light when one considers the typical attitudes companies have towards contractors. Most companies do not partner well with their contractors. Instead, they treat them as disposable entities. If a contractor makes a mistake, the company cancels the contract and hires a new contractor. This attitude makes it difficult for the contractor to partner with the company. If companies today are going to use contractors for maintenance, then they must learn to work closely with their contractors and develop a partnering arrangement.

The partnering arrangement with contractors must be developed to a point at which the contractor feels valued. Many contract firms today believe their technical input to a client company is not valued. In many cases, while doing a maintenance repair, the contract personnel discover other problems. The client company too often assumes that the contractor is just trying to create work, and disregards the contractor's input. In reality, the contractor is trying to save the company money. This example shows that poor partnering with a contractor is an expensive way to do maintenance.

Geographical Organizational Structures

Maintenance organizations may be organized geographically in three basic ways: centralized, areas, and hybrid.

Centralized Organization

In a centralized organization, all personnel report to one central location from which they are directed to work locations. The central organization provides the benefit of more extensive use of the personnel. This better utilization is due to the fact that technicians can always be directed to the highest priority work no matter where its location is in a plant or facility. If properly controlled, a central maintenance organization reduces the amount of nonproductive time for maintenance.

However, the disadvantage of a central organization becomes more noticeable in large plants. The disadvantage is slower response time caused by increased travel time. If there is a problem in one area of the plant and the workers are in other areas of the plant, it takes time to find them, re-deploy them, correct the problem, and then return them to their original assignments.

Organization By Areas

The second organizational scheme focuses on areas. In this scheme, maintenance personnel are assigned to specific areas within a plant or facility. However, a small group of maintenance personnel is always kept in a central location for data collection, analysis, crew scheduling, work planning, etc. In the area configuration, organizations usually respond in a timely manner, because the maintenance personnel are close to the equipment. The disadvantage of an area organization is finding enough work to keep all the maintenance personnel in an area busy. The opposite problem can occur when excessive equipment breakdowns exceed the capabilities of the labor pool within an area. Thus, at one time, one area may have people engaged in lower-level activities, while other areas have equipment breakdowns waiting for personnel. The area concept makes it difficult to move people from one area location to another, due to specialty skills or just distance.

One of the biggest advantages of area organizations is that they help to instill in maintenance workers a sense of ownership of the equipment. In area organizations, the maintenance personnel usually work the same schedule as the operations and production personnel. This schedule allows them to develop better lines of communication with operations and production personnel. Maintenance and production personnel come to understand each other's strengths and weaknesses, and these are taken into consideration during the work cycle. Because both maintenance and production want the equipment to run, they tend to work more closely together to ensure that the equipment does run. The equipment is more likely to be operated correctly and maintained at higher levels than are typically found when maintenance is a centralized organization.

Hybrid Organization

A hybrid organization, or combination organization, is the third option. In a hybrid organization, some maintenance personnel are assigned to areas and the remaining personnel are kept in a central location. The area personnel care for the routine maintenance activities, build relationships with the operations personnel, and develop ownership. The central group supports the area groups during shutdowns, outages, major maintenance, etc.

Which is the best arrangement? The rules of thumb are that central organizations are more effective in smaller, geographically compact plants; area organizations usually perform well in midsize plants; and combination organizations are best for large plants. When developing any maintenance organization, one must give the plant size and organizational geographical structure careful consideration. If one uses the wrong geographical structure, excessive staffing may be required to service the equipment properly. If a central organization is used to service a large plant, the travel time to get to the equipment and the resulting downtime may create havoc, with production schedules constantly disrupted.

Reporting Structures

Another way to look at maintenance organizations is to consider their reporting structures. Maintenance organizations can use a variety of structures, including the maintenance-centric model, the production-centric model, and the engineering-centric model.

The Maintenance-Centric Model

In the maintenance-centric model, maintenance reports to a plant or facility manager at the same level as production and engineering. This model provides a balanced approach, with the concerns of all three organizations weighed equally by the plant manager.

All maintenance personnel in the plant report through a maintenance manager. If the organization is larger, there may be levels of supervisors reporting to the maintenance manager. Maintenance-staff functions, e.g., planners and maintenance engineers, also report to the maintenance manager. Construction and project engineers report to the engineering manager, but no maintenance resources are deployed by engineering. Also, all production or operations personnel report through the production or operations manager, but no maintenance resources are under the control of the production or operations manager. This structure is optimum for organizations learning maintenance controls and philosophies. It is a good structure to start with and it can be developed to support world-class initiatives such as cross-functional teams and operator-based maintenance activities. This organization is shown in Figure 3-4.

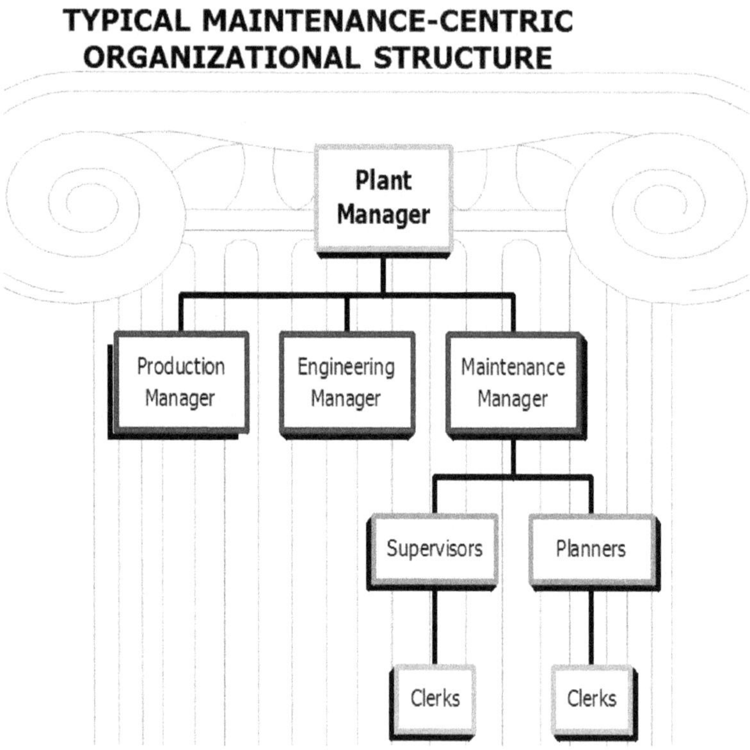

Figure 3-4, Typical Maintenance-Centric Organizational Structure

THE PRODUCTION-CENTRIC MODEL

A second model is the production- or operations-centric organization. In this model, maintenance resources are deployed by the production or operations managers. At first glance, this arrangement might seem to be a good idea. In reality, it rarely works because very few production or operations managers have the necessary technical skills to properly deploy maintenance resources. These cases usually lead to less use of the maintenance work force and, in turn, more

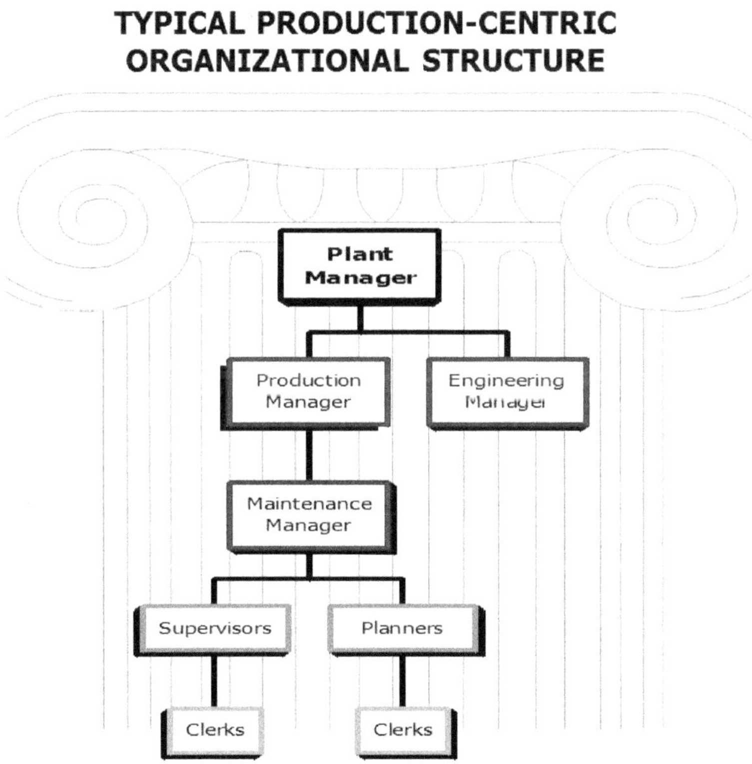

Figure 3-5, Typical Production-Central Organizational Structure

equipment downtime. When maintenance resources report to production or operations, maintenance generally deteriorates into the role of "fire fighting" or "fix it when it breaks." The production-centric model is shown in Figure 3-5.

Another consideration when assessing production-centric reporting structures is the compensation structure for supervisors. In most cases, the production or operations supervisors are rewarded for meeting some level of production throughput or a capacity target-number. Because their compensation is based on output, they have little incentive to perform good maintenance on their equipment. In most cases, under production-centric reporting structures, good maintenance practices are sacrificed to meet production targets. However, if equipment availability or life-cycle costing numbers are included as part of their compensation, then maintenance may be properly managed in this kind of environment.

The Engineering-Centric Model

A third structure commonly found today is an engineering-centric organization. In this structure, maintenance reports to engineering. Thus, construction engineering, project engineering, and maintenance all have the same supervision, e.g., the plant engineer. On the surface, this arrangement appears workable. However, it typically leads to problems. The main problems arise because of projects. Typically, the performance of engineering supervisors is assessed based on their completing projects on time and under budget. If a project gets behind, maintenance resources often are diverted from preventive maintenance and other routine tasks to project work. Although assigning maintenance resources to a project may help complete the project on time, existing equipment may suffer from a lack of maintenance. This structure is shown in Figure 3-6.

TYPICAL ENGINEERING-CENTRIC ORGANIZATIONAL STRUCTURE

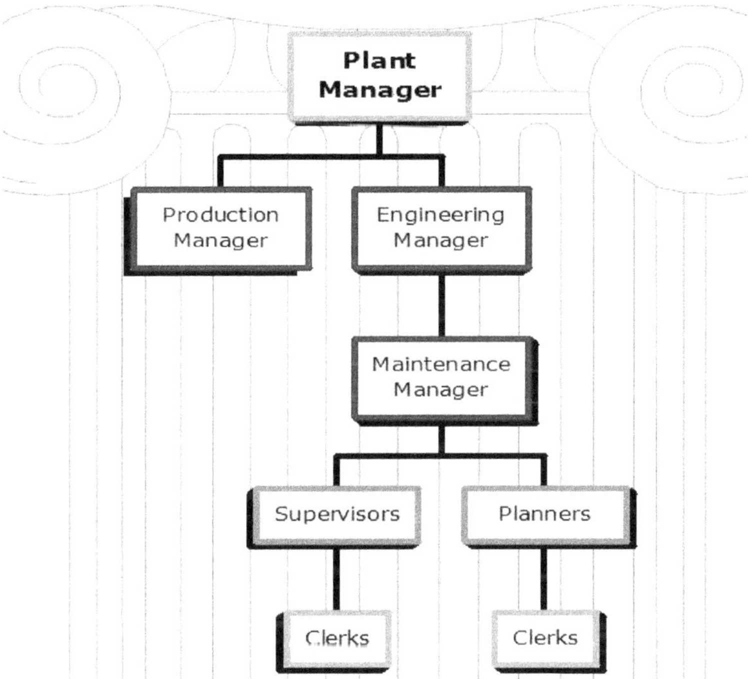

Figure 3-6, Typical Engineering-Centric Organizational Structure

Then, a second, more long-term problem develops. The attitude of the work force is affected. Maintenance personnel enjoy working on projects because all of the equipment is new. Over time, they tend to develop less of a maintenance attitude and more of a project attitude. This shift in attitude leads to their wanting to perform less maintenance work and more replacement work. The maintenance personnel become, in effect, parts replacement specialists rather than maintainers or repairers. This situation can lead to excessive inventory and new equipment purchases.

Whatever the structure of a maintenance organization—and structure does vary from organization to organization—it must have the proper focus. Maintenance is a technical discipline. Maintenance personnel are the stewards of the technology in a plant or facility. If the maintenance organization does not have a technical focus, the assets and equipment will be sub-optimized.

Therefore, if maintenance is sacrificed to achieve short-term production goals or to support engineering construction projects, the maximum return on investment in the existing assets is never achieved. This situation weakens a company's competitive position in its marketplace. If any organizational redesign is proposed for maintenance, both short-term and long-term issues must be examined.

Roles and Responsibilities

In order for maintenance organizations to be effective, certain roles and responsibilities must be defined and assigned. It is beyond the scope of this material to consider all possibilities; still, the following are general guidelines that can be used. Although an organization may not use each of the individual job titles listed in the following section, each of the task lines must be assigned. Thus, an organization may not have a first-line maintenance foreman or supervisor who has a responsibility for each individual line item. Nevertheless, the line-item task descriptions are essential if maintenance is to be managed and, ultimately, the company's assets cared for.

First-Line Maintenance Foremen or Supervisors

The following tasks are typically the responsibility of first-line (or front-line) maintenance foremen or supervisors:

1. Direct the maintenance work force and provides on-site expertise.

When maintenance technicians s are working on an assignment and have questions or need clarification about how to perform a task, first-line maintenance foremen or supervisors should be able to provide guidance. They are also responsible for making individual job assignments and tracking the progress of individual craft assignments.

2. Insure that work is accomplished in a safe and efficient manner.

First-line maintenance foremen or supervisors are responsible for seeing that all technicians for whom they have responsibility work safely and are provided the information, tools, and direction to work efficiently.

3. Review work planning and scheduling with the planner.

First line maintenance foremen or supervisors are responsible for providing feedback to the maintenance planner to ensure that job plans are efficient and effective and that scheduling is accurate.

4. Insure quality of work.

While most maintenance technicians will perform quality work, on occasion they are pressured to take shortcuts. First-line maintenance foremen or supervisors are there to insure they have the proper time to do a quality job the first time.

5. Insure equipment availability is adequate to meet the profit plan.

Quite plainly, this task assigns responsibility for the equipment or asset uptime to first-line maintenance foremen or supervisors.

6. Work with plant or production supervision to insure first-line maintenance is being performed by operators.

If the production or operations group is performing first-line maintenance on their equipment, the first-line maintenance foremen or supervisors have a responsibility to insure the work a) is really being performed, b) is being performed safely, and c) is being performed to the appropriate standards.

7. Verify the qualifications of hourly personnel and recommend training as needed.

When making individual work assignments and observing the maintenance technicians performing these assignments, first-line maintenance foremen or supervisors should be able to observe training needs. As these training needs are identified, it is up to the first-line maintenance foremen or supervisors to see to it that the appropriate training is provided as required.

8. Enforce environmental regulations.

As part of the management team, first-line maintenance foremen or supervisors have the responsibility of insuring that all maintenance technicians observe all environmental regulations. This should include appropriate documentation, work practices, and procedures.

9. Focus downward and are highly visible in the field.

It is a responsibility of the first-line maintenance foremen or supervisors to manage the maintenance technicians at least six hours per day, with no more than two hours per day spent on paperwork or meetings. This is known as the 6/2 rule. It is not cost-effective to have first-line maintenance foremen or supervisors performing clerical paperwork as the major part of their work.

10. Champion proactive maintenance vs. reactive maintenance.

First-line maintenance foremen or supervisors have a responsibility to encourage all production and operations personnel to turn in work for planning and scheduling. Doing this is designed to prevent production and operations personnel from requesting work in a "do it now" or reactive mode; it also helps to insure that maintenance is planned, scheduled, and performed in the most cost-effective manner.

11. Administer the union collective bargaining agreement.

As management representatives, the first-line maintenance foremen or supervisors are responsible for seeing that the components of any collective bargaining agreement are carried out.

12. Monitor the CMMS.

It is also the responsibility of first-line maintenance foremen or supervisors to insure that all data collected by the hourly employees assigned to them are accurate and complete when being entered into the CMMS, if the company uses one.

13. Implement preventive and predictive maintenance programs.

The first-line maintenance foremen or supervisors are responsible for insuring that the craft workers are qualified, and that the crew has the skills necessary to perform the appropriate preventive and predictive maintenance tasks. In addition, the first-line maintenance foremen or supervisors and their crew have a responsibility to improve the preventive and predictive maintenance program constantly. This responsibility may range from improving the individual steps on a preventive maintenance task to implementing new technology for predictive maintenance.

It is not the purpose of this text to determine the job roles and responsibilities for every company. However, each of the thirteen task items just described must be assigned and performed if maintenance is to be properly supervised. The question each organization must ask is: who has the responsibility for each of these task items?

Maintenance Planners

Another role in a maintenance organization is that of maintenance planner. Maintenance planners are different from supervisors or foremen. Whereas supervisors manage the maintenance technicians , planners provide logistic support to them. The following are the typical responsibilities for maintenance planners:

1. Plan, schedule, and coordinate corrective and preventive maintenance activities.

Planners accomplish this task by studying and managing work requests; analyzing job requirements; and determining materials, equipment, and labor needs (such as blueprints, tools, parts, and technicians' skill requirements) in order to complete maintenance economically and efficiently. Maintenance planners are the logistics people. They have the responsibility for removing nonproductive time from the maintenance work force. Their basic responsibility is to insure that when the maintenance work is ready to be performed, there will be no delays during the execution of the work.

2. Develop a weekly schedule and assist the maintenance first-line maintenance foremen or supervisors in determining job priorities.

Planners make changes and adjustments to the schedule and work package after reviewing them with the first-line maintenance foremen or supervisors. Planners maintain a complete and current backlog of work orders. As work is requested, the request is given to the planners. They examine the request, plan the job, and review the job with the foremen or technicians. Once the job is planned and approved, it is placed on the schedule. The planners review the weekly schedule with the first-line maintenance foremen or supervisors before the start of the work week. Recommendations by the first-line maintenance foremen or supervisors that require changes are incorporated into the schedule, and the schedule is then published by the planners.

3. Insure that the CMMS software data files are complete and current.

Planners complete this task by gathering equipment and associated stores information for the entire plant or facility. They develop standardized codes for the equipment, stores, and task craft assignments for all maintenance activities. In short, planners are the keepers of the CMMS software data files. They constantly review information being input into the CMMS for accuracy and completeness.

4. May assist with stores and purchasing functions.

At smaller sites, where planners do not have a full-time work load—typically planning for fewer than 15 to 20 craft workers—planners may assist with stores and purchasing functions. For example, they may be involved in controlling the inventory by ordering, issuing, returning, adjusting, and receiving stores items.

5. Identify, analyze, and review equipment maintenance problems with maintenance engineering.

Planners revise the maintenance management program as necessary to improve and enhance plant and facilities operations. Because planners maintain the work order system, any repetitive problems should be apparent to them. They then review repetitive problems with maintenance engineering to find a resolution. At this level, the resolution typically will be adjustments in the preventive or predictive maintenance program. By adjusting these programs, planners provide a solution to the problem. If the problem is not related to the preventive or predictive maintenance program, then the planners refer it to the maintenance engineer for resolution.

6. Assist in educating operations or facilities personnel in maintenance management.

Because planners are so well versed in maintenance tools and techniques, they should participate in training other company employees in maintenance management fundamentals.

This list highlights the typical responsibilities of a planner. Again, if the organization does not have planners, who, then, is responsible for each of these task items? If maintenance is to be performed economically and efficiently, each of these task items must be assigned. In many organizations, a common mistake is to make the first-line maintenance foremen or supervisors both supervise and plan. However, when first-line maintenance foremen or supervisors have full loads (typically 8 to 12 maintenance technicians), they will not be able to properly supervise and plan. Because they cannot do both jobs correctly, maintenance will not be performed as efficiently and effectively as it could be

Up to this point, the focus has been on managing the maintenance work force and providing the support needed to make them efficient and effective. Now the transition is made to managing assets or equipment. If the first two task lists are properly assigned and completed, then the organization is collecting data through the work order system and the CMMS. The next task list entails making this data effective in the maintenance of management.

Maintenance Engineers

The following tasks are typically the responsibility of the maintenance engineers:

1. Insure that equipment is properly designed, selected, and installed based on a life-cycle philosophy.

Many companies today consistently purchase equipment based on the low bid. Quite simply, if they are not performing the tasks listed for the maintenance foremen and maintenance planners, the company lacks the data to purchase equipment based on the life-cycle philosophy. Without the data, the purchasing and accounting departments will purchase the lowest cost items, which may or may not be the best long-term decision. Thus, collecting maintenance-cost data is important.

2. Insure that equipment is performing effectively and efficiently.

This task is different from tracking uptime. It means insuring that the equipment, when it is running, is at design speed and capacity. When focusing only on maintenance, many companies set goals in terms of uptime. However, many companies do not realize, as they move into this aspect of maintenance, that the equipment may be running at only 50 to 60% of capacity. Thus, understanding design capacity and speed ultimately is more important than measuring uptime.

3. Establish and monitor programs for critical equipment analysis and condition monitoring techniques.

Maintenance engineers are responsible for insuring that the appropriate monitoring techniques are used for determining equipment conditions. This information is then given to the planners so that effective overhaul schedules can be determined. These techniques should also help eliminate unplanned maintenance downtime.

4. Review deficiencies noted during corrective maintenance.

As mentioned in task #5 for the maintenance planners, the engineers and the planners periodically review equipment maintenance records. If they observe continual problems with equipment, and the problems are not with the preventive or predictive maintenance programs, then the maintenance engineers will be responsible for finding solutions to the problems.

5. Provide technical guidance for CMMS.

Maintenance engineers review the data in the CMMS. They make recommendations about the types of data and the amount of data being collected. They may also recommend problem, cause, and action codes for properly tracking maintenance activities.

6. Maintain and advise on the use and disposition of stock items, surplus items, and rental equipment.

Maintenance engineers review spare parts policies for plant equipment. This review is to insure that the right parts are in stock—in the right amounts.

7. Promote equipment standardization.

Maintenance engineers help insure that the company is purchasing standardized equipment. Standardiza-tion reduces the number of spare parts required and the amount of training necessary. It also reduces the overall maintenance budget. Standardiza-tion requires data from the CMMS. If the organization is not collecting data through the maintenance foremen and maintenance planners, then maintenance engineers will not have the data required to implement equipment standardization.

8. Consult with maintenance craft workers on technical problems.

Maintenance engineers consult at a technical level with maintenance technicians concerning equipment or work-related problems. This consultation may be about advanced troubleshooting or even equipment redesigns.

9. Monitor new tools and technology.

Maintenance engineers are responsible for staying abreast of all the tools and technology that are available in the maintenance marketplace. Therefore, they are responsible for reading books and magazines, attending conferences, and interfacing with other maintenance engineers to gather this data.

10. Monitor shop qualifications and quality standards for outside contractors.

Maintenance engineers are responsible for insuring that all outside contractors are qualified and that the work performed by the contractors is of the proper quality.

11. Develop standards for major maintenance overhauls and outages.

Maintenance engineers are responsible for examining outage and overhaul plans for completeness and accuracy. They then make appropriate recommendations to the planners for adjustments in the plans or schedules.

12. Make cost-benefit reviews of the maintenance programs.

Periodically, maintenance engineers review maintenance programs for their areas of responsibility and determine whether the work should be performed by operators, maintenance technicians , or outside contractors. In addition, they review what work needs to be done, what work can be eliminated, and what new work needs to be identified and added to the maintenance plan.

13. Provide technical guidance for the preventive and predictive maintenance programs.

Maintenance engineers periodically review the preventive and predictive maintenance programs to insure the proper tools and technologies are being applied. This review is typically in conjunction with the maintenance planners.

14. Monitor the competition's activities in maintnance management.

Maintenance engineers are responsible for gathering information about competitor's maintenance programs. This information may come from conferences, magazine articles, or peer-to-peer interfacing and should be reviewed for ideas for potential improvements in their company's maintenance program.

15. Serve as the focal point for monitoring performance indicators for maintenance management.

Maintenance engineers are responsible for developing performance indicators for maintenance and reviewing those with the maintenance managers.

16. Optimize maintenance strategies.

Maintenance engineers are responsible for examining maintenance strategies and ensuring that they all are cost effective.

17. Responsible for analyzing equipment operating data.

Maintenance engineers insure that equipment is operating as close to design parameters as possible. Doing this insures that there is no wasted production from less-than-optimal equipment capacity.

In brief, maintenance engineers are responsible for properly managing assets. They are key individuals if companies are going to maximize asset utilization. Maintenance engineer are different from project engineers. Whereas project engineers concentrate on new construction and new equipment, maintenance engineers concentrate on optimizing existing equipment or assets. Ultimately, it is the maintenance engineers' goal to insure that their company gets as much or more production from its assets than any other company does that has the same kinds of assets.

Maintenance Managers

The following list describes the tasks for the maintenance managers, or the individuals responsible for managing all of the maintenance functions for a company:

1. Responsible for the entire maintenance function, including the planning, supervising, and engineering staffs.

These individuals have the responsibility for all maintenance activities within the company. The maintenance planners, supervisors, and maintenance engineers report directly to these individuals. This structure produces one-point accountability for the entire maintenance program.

2. Coordinate closely with counterparts in other in-house organizations.

Maintenance managers coordinate with other organizations to insure that company objectives are being met. They communicate closely with production or operations, project or construction engineering, accounting, purchasing, and other organizations. As a result, the organization maintains its focus on optimizing the company's assets.

3. Promote proper understanding of the maintenance function to other organizations.

Maintenance managers educate other organizations within the company regarding the value of maintenance management. This education is intended to help other organizations understand the impact that their functions have on the maintenance organization's efforts to properly maintain the company's assets.

4. Insure that all supervisors, planners, technicians, and maintenance engineers are properly educated and trained.

To be able to fulfill their responsibilities, other maintenance personnel need to be educated and trained. Insuring that education and training takes place is one of the most overlooked responsibilities of maintenance managers. Technology is constantly changing. The entire maintenance organization's skills must be kept up-to-date if it is to fulfill its responsibilities correctly.

5. Take responsibility for planning, cost control, union activities, vacation planning, etc.

Maintenance managers are responsible for all the logistics and personnel activities for the maintenance organization. They also administer the maintenance budget and insure that the maintenance function meets its budgetary requirements.

6. Has responsibility for delegating assignments to the appropriate personnel.

Maintenance managers have a responsibility to insure that the appropriate personnel are in the proper staff positions within the organization. In other words, managers have the responsibility to see that the organization is staffed correctly and operates smoothly.

Maintenance Organization and Staffing

In the current climate of downsizing, organization and staffing are among the most critical issues affecting maintenance. How is the maintenance organization staffed? While companies have tried many different staffing formulas over the years, the only perennially successful one is staffing the maintenance department based on work backlog. A maintenance work backlog is the amount of work currently identified as needing to be performed by the maintenance department. This amount of work is measured in hours. Many have tried to measure backlog by the number of work orders, percentage of production hours, etc., but it never works. The only true measure of backlog is based on hours of work to be done. To calculate the backlog, in addition to knowing the hours of maintenance work needed, it is also necessary to understand current work force capacity.

The formula for calculating backlog is as follows:

Backlog = identified work in hours ÷ craft capacity per week (in hours)

For example, a backlog contains 2,800 hours of work that is currently identified. The current work force has ten technicians who each work 40 hours per week plus 8 hours of overtime per week. Total hours worked per week by the technicians, then, is 480 hours. The company also uses two outside contractors for 40 hours each per week—another 80 hours. Therefore, the total capacity for the work force is 560 hours. If the 2,800 hours in the backlog is divided by the 560 hours of capacity, this produces a backlog of five weeks. An optimum backlog is considered to be between two and four weeks of work.

At first glance, the five-week backlog does not seem to be too far from the optimum. If, however, an organization scheduled 560 hours of work from the backlog for their crews next week, it would be virtually impossible to accomplish that 560 hours of planned work. The reason is the amount of emergency or reactive work that occurs on a weekly basis. In some companies, emergency and reactive work makes up as much as 50% of the maintenance department's work-allocation each week. If this is the case, then only 280 hours of additional work can be done.

In addition, the technicians have routine assignments—lube routes, re-builds, and other routine activities. There are also meetings, absenteeism, vacations, and training. When all of these factors are considered, the actual hours available to be scheduled might be about 200. If only 200 hours are available to be scheduled, then the backlog is actually 14 weeks, an unacceptable amount. One can only imagine the reaction of the production department when it submits a work order that it expects to be done within two to four weeks and is told it may take as long as three-and-a-half months to complete.

Although this scenario is bleak, there is a second, more important problem. That problem is the proper identification of work that needs to be performed by the maintenance organization. The maintenance department is staffed based on identified, not actual work. For example, if someone today performed an equipment walk-down throughout your entire plant, how much work could be identified that needs to be done, but is not yet recorded? Hundreds, if not thousands, of hours of work may need to be performed. This unrecorded work, along with the previously described factors, lead to underestimating the backlog and, ultimately, to insufficient staffing of the maintenance department. The organization would revert to a reactive condition because current staff can never accomplish the required work in a proactive mode.

Another common practice in industry further compounds the problem. This practice is what is identified by many companies as a backlog purge. A backlog purge occurs when all small jobs are removed from the backlog or deferred to another time. The jobs are those perceived to be noncritical and to be done at another time. This is a mistake! Work should be identified and performed before it becomes critical. The attitude is "It's only a small job, not to be worried about." However, over time, small jobs become big jobs. In reality, then, the organization is saying, "We only want to work on big jobs," or "We will wait till it becomes a critical problem before we address it."

Backlog purges are used by companies to justify downsizing or reductions in staff. It would be quite unusual for any company to defer or cancel small orders from customers and make the customers wait until after the company fills the big orders before accepting and running the small orders. The parallel with maintenance is clear. If a work order is turned in, approved, and put in the backlog, then it is a legitimate request. It should never be canceled or deferred until it becomes an emergency.

The goal should be to maintain the backlog in the two-to-four-week range. If the backlog begins to increase or trend above four weeks, then more resources should be added. From the formula, one can see that there are three options for resources. A company can con-

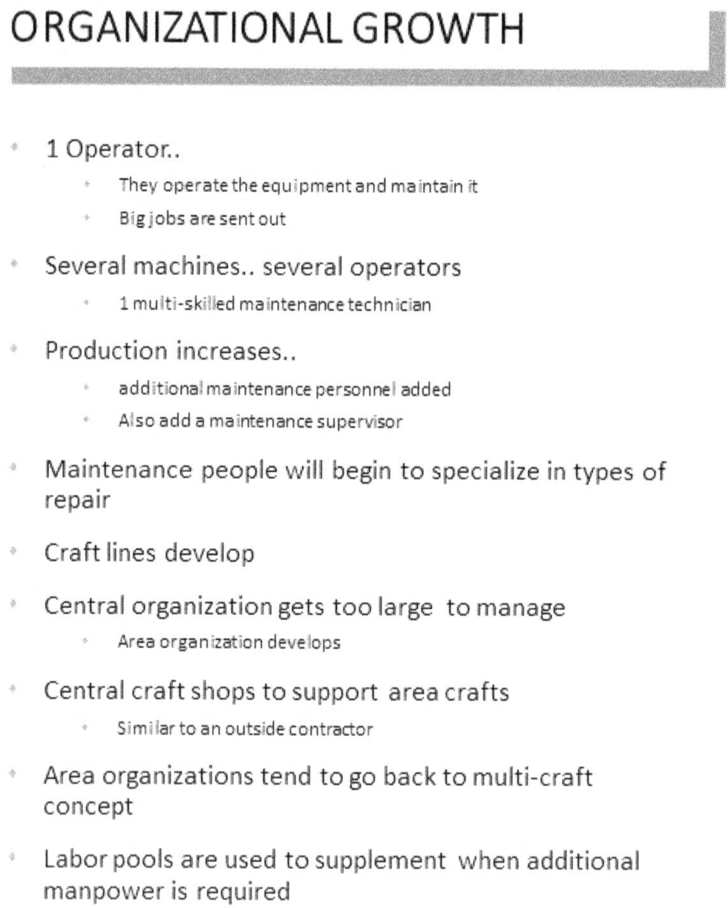

tract out more work, its employees can work more overtime, or it can hire more employees. Conversely, if the backlog begins to trend or drop below two weeks, the company needs to reduce resources. The company could reduce the amount of outside contract work, reduce the amount of craft overtime, or ultimately reduce the size of the maintenance work force. If the backlog is calculated weekly and tracked annually, seasonal trends and other spikes can be clearly seen. By reviewing these types of records, a manager can insure that the department is properly staffed.

Organization Evolution

In examining maintenance organizations, we find they all have the same growth pattern (see Figure 3-7on the following page). When companies are small, they may only have one machine. The operator of the machine runs and maintains it, performing any small repairs or services. If a large breakdown occurs, the machine is disassembled by the operator and the defective parts are sent out for repair. As the company continues to grow, several machines are added. Growth also necessitates adding several production workers. Because the production workers are not dependent on one machine any longer, the first maintenance worker is added. This individual will be multi-skilled in order to care for the variety of repairs that will be needed.

The third step in the growth pattern is the addition of more machines and production workers. This step leads to the addition of more maintenance workers. With this new level of manpower required by maintenance, it is no longer convenient to have the maintenance workers report to the production supervisor. Instead, it is necessary to have a maintenance supervisor in place. The fourth step in the growth pattern is to watch the maintenance personnel begin to specialize in their particular skill areas. Craft technicians become proficient at repairing a particular piece of equipment, or a particular type of repair. As the number of craft technicians increases, the specialization continues. The fifth step in the growth pattern is the development of craft lines. This development may be due to union influence, or simply the natural progression of step four. The lines can be either strict or informal, but they will become increasingly distinct.

The sixth step in the growth pattern occurs when the organization becomes too large to manage from a central location. Several factors may contribute to the management problem. The internal geography of the plant is one factor. For example, if the plant covers several hundred acres, it may be physically impossible to manage it from one location, even with the help of radios, bicycles, or manned carts. At this stage, the organization may divide into the area concept, allowing small maintenance departments, such as those described in the first three steps.

Most organizations develop two alternatives at this stage: further internal growth or outside contracting. Internal growth will develop central crafts or shops to support the area organizations. Thus, as in step one, when the repair is too large, complicated, or requires special equipment, it is sent to the central shops. As more work is required from the central shops, they tend to grow, whereas the area organizations tend to add employees only when new demands are made on their area (such as new equipment additions) or when attrition occurs. Outside contracting occurs when the company either does not have the resources to implement central crafts or decides it is more cost effective to contract with a local shop for machining, rewinding, or installation. The determining factors here are skill level of the contractor's workforce, response time, and synergism between the contractor and the company.

In the final growth step, the area organizations tend to go back to the multi-craft concept, allowing for the maximum flexibility of the labor resources assigned to an area. To assist in the peak work periods, the central organization might maintain a pool of qualified individuals, capable of being proficient in various areas of the plant. There are companies with as many as 30 or 40 area organizations within a single plant, coupled with central organizations and outside contracting, managed by area organizations, reporting to a central maintenance management organization, that provide an optimum service/cost factor arrangement. Each company will make the policy decision several times before they find its optimum organization.

How management perceives maintenance, how maintenance perceives its role, and the attitudes of managers and craft technicians toward one another are also important. The attitude management shows toward the maintenance craft technicians helps to establish pride in workmanship. If an organization is in a "fire-fighting" mode—strictly fixing it as quickly as it can be fixed—then workmanship suffers. Craft technicians will get into a habit of fixing it to "just get by."

When this environment begins to change to a more proactive environment, the craft technicians have difficulty adjusting to the "fix-it-right-the-first-time" attitude. It is the same problem anyone has when trying to break habits that have become entrenched. The in-fighting between maintenance on the one side and production, operations, and facilities on the other also must end. If maintenance is to contribute to the overall profitability of the corporation, all parts of the organization must be given responsibility and accountability. Areas where attitude toward maintenance manifests include:

— Maintenance shop locations
— Maintenance equipment
— Maintenance incentive programs

Maintenance repair shops should be located in areas convenient to the job locations. It should be easy for rebuildable items, repairs, and other maintenance work to be brought into the shop area, where the larger tools are located. Therefore, the area must have adequate clearance for fork lifts, overhead cranes, and other transportation methods. Repair shops should also be located in areas of the plant where excessive noise levels do not make working in the area difficult. For example, in one plant, the maintenance shop was located beside the plant's rock crushers. The noise level made work without hearing protection impossible. In addition, the lathes for the maintenance shop were nearby; the vibration made any finishes smoother than .010 impossible. Needless to say, the maintenance organization did not have a high sense of self-worth.

The equipment in the maintenance shop is also very important. The quality of the tools helps to determine the quality of the work performed by the shop. If the maintenance department does not have the tools and equipment needed to maintain plant equipment, one can hardly say they do not do their jobs. For example, it is difficult to maintain solid state control equipment with a VOM multimeter instead of an oscilloscope. Asking the maintenance personnel to produce precision work with old, worn out tools and equipment reflects a measure of their importance. Maintenance incentive programs are not properly utilized in most plants and facilities to produce motivated craftsmen. Incentive programs can be tied to uptime, production rates, or total departmental operation for the purpose of motivating the maintenance workforce. If the maintenance personnel believe that they can increase their financial status by increasing performance, they also can work more productively.

Summary

The organization for maintenance can be varied and adjusted to fit many circumstances. The options detailed in this chapter are used by various organizations around the world. The main points to remember are these:

1. All organizations exist to accomplish certain goals or objectives. Maintenance is no different; be sure yours are known and accepted.

2. Organizing the maintenance function is important. Incorrectly organizing the resources can result in excessive maintenance costs.

3. Contractors are being increasingly used in the maintenance environment. Careful policy decisions can make contractors cost effective.

4. Attitudes toward maintenance are shown by the way the organization is treated when it comes time to dedicate re sources. Always insure that maintenance has proper tools, proper locations, and incentive to work.

The previous chapter on benchmarking looked at the hidden narrative or enablers that do not show up in pure statistics. Organizational structure and discipline is one of the benchmarking enablers. The ISO-55000 standard also requires an analysis of the organization to insure it supports asset management. The organizational focus, structure, and disciplines discussed in this chapter are difficult points to analyze when benchmarking. They are seldom explored to the depth necessary to provide an understanding of how they impact the ability of an organization to achieve Best Practices in maintenance. If a company is to truly understand a benchmarking partner's practices, then the areas discussed in this chapter must be thoroughly understood.

Chapter 4

Maintenance Training

Technical training has been called one of the largest weaknesses of the present maintenance structure in the United States.

If you think education is expensive;
Try to count the cost of ignorance

Training Expenditures

Estimates suggest that a company should budget training dollars for its technical personnel on an annual basis. Several methods can be used to track the training expenses. The first is expenditures per employee. A recent survey (2013) by the ASTD (American Society of Training and Development, www.astd.org) showed that companies average from $700 to $1,800 per employee per year. A second method of tracking training expenses is by average hours . The survey showed an average of 30.3 hours . The survey figures are summarized in Figure 4-1.

The problem with using the averages is that they don't actually show what type of training is being conducted. For example, a further

FIGURE 4-1 TRAINING STATISTICS: AVERAGES
▶ Expenditures per Employee: $1195 $700 low ◀——————▶ high $1800
▶ Hours per Employee: 30.3
▶ Employees per Trainer: 299:1 97:1 low ◀——————▶ high 400:1
▶ Percentage of Expenditures to Outside Firms: 28% Tuition reimbursement: 11%
▶ Classroom Training: 55% of Total Training
▶ Technology Training: 20% of Total Training

breakdown of the training expenditures shows that the technical training comprised only 20% of the total. This amount is insufficient to maintain the technical skills of the technicians. The lack of technical skills will restrict the deployment of maintenance resources and increase the overall cost of maintenance. In addition, 27% of respondents to the ASTD's 2013 survey said that informal learning accounted for over 50% of all of the learning that took place in their organization. This type of learning is insufficient to pass on the technical skills necessary to be successful in maintenance organizations.

The Value of Training

Consider the statement in Figure 4-2.

This statement provides a realistic illustration of what happens in maintenance departments today. When the employees do not have the proper skills, managers defer work until someone is scheduled who they believe is competent enough to perform the job. This deferral results in work delays, damage to equipment, and expressions of dis-

FIGURE 4-2 TRAINING VALUE

With a good crew,
one would make the inspections;
with a poor crew, one would rather
take a calculated risk of failure
up to a certain level of severity.

satisfaction from the operations or facilities managers. This scenario highlights why having an appropriately trained technical workforce is so important. However, even if a company believes that its workforce is technically competent, Figure 4-3 shows why ongoing training is important.

If a company modernizes at an accelerated rate, the time frame in Figure 4-3 is shortened even more. Reflecting on the technological advances that have been made in plant and facilities equipment over

FIGURE 4-3
IMPORTANCE OF TRAINING

Eighty percent of the skills
of those now working
in technical areas
will be obsolete
in three to five years.

FIGURE 4-4 TRAINING COSTS VS. SAVINGS
▶ When we first tracked costs for maintenance bearings, we found that just by training 120 people in bearings, we saved $4.5M in one year.
▶ Training is a hedge against obsolescence. We need to look at employees as assets, rather than as costs.
▶ We think nothing of spending $130M on equipment, but if you walk in and say you are going to spend $2M on training, people look at you like you are crazy.

the past twenty years, we can quickly see the truth in these statements. How can comprehensive technical training be cost justified in tight economic times? Consider the point made in Figure 4-4.

This quote is by a manager who clearly understands the value of training. If this type of return on investment is presented with focused

FIGURE 4-5 OPERATOR TRAINING: LIFT TRUCKS
▶ Maintenance costs ran 14% higher for companies with untrained operators.
▶ One company showed a 20% reduction in maintenance costs after an extensive training program.
▶ For a company with $10M in sales, a 10% reduction in operating costs produced more pretax profit than a 10% increase in sales.

FIGURE 4-6 **MAINTENANCE TRAINING OPTIONS**
▶ Hire Trained Personnel
▶ Vocational Schools
▶ Train In-House
▶ Train at Vocational Schools ♦ (Community Colleges)
▶ Vendor Training
▶ Colleges and Universities
▶ Continuing Education Programs

technical training, then cost justifying technical training is easy in almost any economic environment. Unfortunately, not too many companies use focused training programs. Most herd a group of people in a room, lecture and demonstrate the training, then let the attendees leave, hoping they learned something. This is often termed the "Sheep Dip" training method.

In addition to maintenance training, the training of operations personnel can have a dramatic impact on the cost of maintenance and availability of equipment. Figure 4-5 highlights the value of operator training.

Training Programs

Every company's self-evaluation should consider when formal training programs were last offered for its craft technicians, planners, and supervisors. Without good quality training programs, a maintenance organization will never be cost effective.

Craft Training Programs

Figure 4-6 highlights seven options for overcoming skills shortages in the workforce.

Training Options

Hiring trained personnel can be a quick fix if a company has a severe skills shortage due to the retirement or departure of a key employee. However, it is rarely the cure to a long-term skills shortage problem. First, when highly skilled technicians are hired, they will be expensive. In order to get someone who already has the right skill level, the current pay scale for the craft technicians may have to be revised upward, sometimes considerably.

There are companies that have made maintenance technician positions salary rather than hourly, just to be able to hire the right person. Additionally, there are companies that are paying signing bonuses to insure the right person is hired. Furthermore, employees who join a company just for financial benefits often move to other companies once they get a better offer elsewhere. The company can actually find itself with an even worse skills shortage, especially if proper backup personnel are not mentored.

Vocational schools offer varying degrees of benefits, depending on how well the schools work with industry. The same is true of community colleges. There are numerous examples of joint ventures where the vocational schools or community colleges effectively work with local industries to identify specific training needs. Courses are then developed that address these specific needs. In many cases, the courses are then expanded and offered to all local industries.

A second option using vocational schools and community colleges is to provide training at their locations. The company develops its own course, and for a set fee uses the facilities (lab and classrooms) at the school. This option is typically used when some form of proprietary material needs to be taught. In this case, the company usually furnishes the instructor.

Vendor training programs are often offered by the vendors of equipment or equipment components. The courses are usually supported with good customized materials and the vendors typically provide high-quality instructors. However, the course material is usually specific to only one component or piece of equipment. Furthermore, vendor training tends to include a sales perspective. If these limitations are understood and can be minimized in advance, great value can be derived from these courses.

Four-year colleges and universities often offer higher level technical courses, but most of these are focused on designing equipment, rather than maintaining the equipment. Occasionally universities partner with industries to develop accredited classes, especially in electronics, in which the course content is more focused on maintenance and troubleshooting than on system design.

An extension of this arrangement is found with continuing education programs. These are courses that are developed by subject matter experts, internal or external to the college or university. These programs are marketed by the school. They typically run one to three days and the attendees are awarded the appropriate Continuing Education Units (CEUs). The programs range in scope from very technical orientation to managerial sessions. They are usually offered in an in-house format, thereby lowering the total cost, particularly if numerous individuals need to attend the course.

In-house training is the most expensive option, but also the most effective if properly developed and conducted. There are several levels of in house training programs, discussed below.

Levels of In-House Training

Apprentice Training

The first level of training is the apprentice training program. This level of training takes regular, unskilled workers and gives them the training they need to become skilled craft technicians. The training program combines on-the-job training with classroom training. Most good programs will run three to four years and include hands-on lab sessions.

Some companies work with local vocational schools to fill entry-level craft positions. This arrangement allows the company to specify some of the material that must be covered in the program. The vocational schools benefit from the assistance they get placing their students once they complete the program.

Correspondence Programs

Correspondence programs are a second option for training entry-level craft technicians. These are completed by craft technicians at their own pace and are among the most economical methods used in maintenance training.

Closely related are the "canned" programs that contain information related to each craft line, whether millwrights, electricians, mechanics, etc. The disadvantage of these programs is their generic material, dealing with theory, not real world experience. Although this approach is good for engineering-level training, it is not satisfactory for craft-level training. A second problem is the lack of a knowledgeable instructor to help the apprentice understand and apply the material being studied.

The objection about the material being generic is true in most cases. However, several vendors have programs that specifically address the repair and troubleshooting of components. These programs have been successful and are seen by craftspeople as beneficial. In fact, some journeymen, after seeing the quality of the material the apprentices were using, even bought their own set of materials for reference.

Internal and External Training Materials

Another option is to have training materials developed either internally or by an outside firm that specializes in maintenance training materials. These materials are developed from a job needs analysis, a study that determines just what individuals need to know in order to properly perform their specific jobs. The materials are then written and illustrated from this needs analysis. Because all the materials apply specifically to the craft technician's job, they are readily accepted. The greatest obstacle to using this type of material is its cost. Performing a job needs analysis is by itself expensive. In addition, there are the costs of writing the text, preparing the artwork, and printing and binding the materials. However, the advantage of having a job-specific training program, which apprentices can take at their own pace, can be worth the cost of such a program.

Training Center

The best though most expensive option is to establish a craft training center within the company. It should have the classroom, lab, and real world setting necessary to produce qualified, competent craft technicians. One such program in existence in the early 1970s became a corporate model for one steelmaker. The program structure was designed for a "super-craft" environment, requiring proficiency in mechanical, electrical, and fluid power. The students were required to complete 1040 hours of classroom training, in addition to 3 three years of on-the-job training.

The program took different forms depending on the economic situation. The apprentices attended class either one day per week, one week per month, or one month on and month off. The optimum schedule appeared to be the one week per month format. It allowed the instructors to teach three weeks each month with the remaining week for preparing the next cycle. Five classrooms were set up in which the students sat two to a table. Metal top tables were used so that instructors could bring a display or lab project into the classroom. There were six labs: three mechanical, one fluid power, and two electrical (one D.C. control and one A.C. control). The curriculum consisted of the following studies:

- 260 hours D.C. Electricity
- 260 hours A.C. Electricity
- 260 hours Mechanics
- 260 hours Fluid Power

The course outlines were not theory intensive. Instead, they concentrated on the maintenance and troubleshooting of the components and circuits. The apprentices received just enough design theory to understand why certain designs were used, although not enough to do design. For example, in the mechanics class dealing with V-belts, apprentices learned enough theory to understand speed differentials and forces generated by the rotation. The thrust of the discussion was how these forces create problems in maintaining the V-belts: why the tension requirements are critical, why alignment is important, why the bearing adjustment is important, and why the condition of the sheaves is important. These points made the material applicable to their jobs.

What was the single most important factor in the program? Ultimately, the instructors. They spelled its success or failure. If the instructors could not relate the theory to the real-world needs of the apprentices, the course would be a waste of time. For this reason, the instructors were themselves journeyman craft technicians with experience. They were required to have good presentation skills, to be able to develop logical, coherent outlines, and to develop tests for their materials. Without use of instructors who were respected for their job skills by their peers, the program would have been a failure. The training needed to be conducted in conjunction with on-the-job training also. This aspect was the hardest part of the program, insuring that if apprentices were clearing hydraulic circuits, they had a chance to work on some hydraulic circuits that month. Good communication with the apprentice's department heads and supervisors generally would help in coordination. The reason for presenting this illustration is not to paint a pie-in-the-sky picture for anyone, but rather to show the importance that some companies give to craft training. If the effort to obtain a fully-trained work force is not made, the company cannot expect to achieve a high level of skilled service from its maintenance workforce.

Journeyman Training

Journeyman training is usually related to specific tasks or equipment maintenance procedures. Journeyman training courses can be conducted by in-house experts, vendor specialists, or outside consultants. The training may address a new technology that is being brought into the plant. For example, when vibration analysis was first being introduced into the maintenance environment, there were extensive training programs in the use of vibration analyzers offered by the vendors and consultants. These programs were addressed to the journeyman-level craft technicians who would be involved in the programs.

When new equipment is purchased and installed in the plant, training programs are performed by the vendor on the care and maintenance and troubleshooting of the equipment. Again, it is the journeyman-level craft technicians involved in the programs. Good journeyman craft training programs should be a part of any complete maintenance training program.

Cross Training or Pay for Knowledge

Cross training is becoming increasingly common in progressive organizations. It is a sensitive subject, however, because it generally involves negotiations with the union representation. This type of program is essential if U.S. maintenance costs are going to be brought in line with the costs incurred by overseas competition. The cost savings for the company is found in planning and scheduling the maintenance activities. For example, consider a pump motor change out. In a strict union environment, it would take:

- A pipe fitter to disconnect the piping
- An electrician to unwire the motor
- A millwright to remove the motor
- A utility person to move the motor to the repair area

The installation would go as follows:

- A utility person to bring the motor to the job area
- A millwright to install the motor
- An electrician to wire the motor
- A machinist to align the motor
- A pipe fitter to connect the piping

As can be seen, not only are many people involved in the pump motor change out, but also the coordination to insure that all crafts are available when needed without delay will become extremely difficult. In a multi-skilled or cross-trained environment, only one or two craft technicians would be sent to the job to complete all the job tasks. The advantages of costs and coordination are obvious. But what are the advantages for the employees?

First, the action helps to insure maintenance is a profit center. This means not only being cost effective, but also running the department as a business. For example, picture the following production scenario:

- One person transports material to the job site
- One person inserts the part in the machine
- One person drills the hole
- One person finishes the hole to specifications
- One person machines the finish
- One person takes the part out of the machine
- One person moves the part to the next process

This would not be a well-managed, cost-effective operation. Should maintenance be any different? The goal should be the same: to maximize the utilization of all assigned resources. The employees should work to contribute to the profitability of the corporation; our overseas competitor's employees do.

A second advantage of cross training is that employees have additional skills that insure they are valuable contributors to the corporation's goals. This increases their self-esteem and value. But most of all, any cross-training effort must have financial rewards to the employees. Because cross-trained employees are more valuable to the company, they should receive a higher rate of pay.

For example:

- Apprentice — pay level 1
- Journeyman (one craft) — pay level 2
- Journeyman (1 craft, apprentice 1 craft) — pay level 3
- Journeyman (1 craft, apprentice 2 craft) — pay level 4
- Journeyman (2 craft, apprentice 1 craft) — pay level 5

The list could go on, depending on the crafts involved.

These types of programs are growing increasingly popular in the United States, as they must. If we are to take advantage of potential cost savings, cross training is one of the most important areas to investigate.

Planner Training

Maintenance planners should come from craft technicians who have a good aptitude for logistics. However, planners need training in subjects beyond the skills required by craft technicians. These subjects include:

- Maintenance priorities
- Maintenance reporting
- Project management
- nventory management
- Scheduling techniques
- Computer basics

Without such training, achieving the level of proficiency necessary for a successful planning and scheduling program will be difficult. Training is one of the most important factors in the development of a good maintenance planner. The sources for training materials are similar to those for maintenance craft training materials, and include:

- Correspondence courses
- University sponsored seminars
- Public seminars
- Maintenance consultants
- Maintenance software vendors

The content of the training program for maintenance planners will vary depending on whether the organization is a facility, process industry, food service, vehicle fleet, etc. Each has its own unique needs, even though many of the basic principles are the same. Good, competent, skilled maintenance planners will pay dividends for the investment in their training many times over.

Supervisor Training

Front-line maintenance supervisor positions are filled from craft or planner promotions 70% of the time. They are then familiar with the assignments that they will be responsible for supervising. In a personal observation, rarely does an engineer or another staff person make the transition to becoming a front-line maintenance supervisor. The lack of craft knowledge or experience cannot be compensated for in actual job supervision. One of the major mistakes in selecting front-line maintenance supervisors is promoting them from the craft group without proper training, almost as if management expects them to pick up their management skills by osmosis. Good supervisor training programs should be implemented before supervisory responsibilities are assumed. Some areas that should be addressed in these programs are:

- Time management
- Project management
- Maintenance management
- Management by objectives

The support and understanding of the front-line supervisor helps to determine the success or failure of many programs implemented by upper management. Therefore, supervisors should receive the training necessary to insure their success so that, in turn, their success carries on to the rest of the organization.

Final Points

Training is one of the hidden factors that must be carefully analyzed in benchmarking. Unless the skill set of the workforce is clearly understood, the way in which certain Best Practices are developed, implemented, and sustained will not be understood.

Training is vitally important to all levels of the maintenance organization. Unfortunately in a volatile financial environment, maintenance and training programs are among the first to be cut back. This is especially true of programs that are perceived by management as non-essential, which in most organizations includes training. This short-sighted management philosophy must change if maintenance is to be managed successfully.

Chapter 5

Work Order Systems

Work order systems are one of the keys for successful maintenance management. Work orders, which are the documents used to collect all necessary maintenance information, can be described in many different ways. For the purpose of this text, we will use the following definition:

A work order is a request that has been screened by a planner, who has decided the work request is necessary and has determined what resources are required to perform the work.

Who Uses Work Orders?

Work orders should not be implemented by just the maintenance department, without regard for other parts of the organization. Figure 5-1 lists the groups that should be involved in the use of a maintenance work order system.

FIGURE 5-1 WORK ORDERS
Work orders should satisfy and be used by:
▶ Maintenance
▶ Operations/Facilities
▶ Engineering
▶ Inventory/Purchasing
▶ Accounting
▶ Upper Management

MAINTENANCE

Maintenance is the primary user of the work order. Maintenance requires information such as:

- What equipment needs work performed
- What resources are required
- A description of the work
- Priority of the work
- Date needed by

Other information may be required, depending on the type of facility or plant in which the work order system is used. The main point is that the maintenance organization must get the information needed for good management decisions. If the information cannot be obtained from the work order, it is unlikely reliable information will ever be available from another source.

Operations or Facilities

Operations or facilities also needs input into the work order process. They must be able to request work from maintenance in an easy process. If they have to fill out 15 forms in triplicate, they will be unlikely to participate in the use of the work order, thereby eliminating its effectiveness. Whether it is manual or computerized, the work order system must be easy for operations and facilities personnel to use. They should only be required to fill in brief information, such as:

- What equipment needs work
- Brief description of the request
- Date needed
- Requestor

This information can then be used by the planner to complete the work request and, in turn, convert it to a work order. In some organizations, particularly those moving toward operator driven reliability (ODR), the operations personnel may have as much involvement with the work order system as the maintenance technicians. However, this is still an area for development in most companies.

Maintenance / Reliability Engineering

Because maintenance / reliability engineers are usually charged with the effectiveness of the preventive and predictive maintenance programs, they need input into the work order system. In addition to requesting work for engineering services, maintenance / reliability engineers also need access to historical information. If accurate and properly maintained, historical information can help engineers operate a cost effective preventive maintenance program. Without accurate information, the PM and PDM programs become guesswork. Therefore, the maintenance / reliability engineering staff will need information such as:

- Mean Time Between Failure
- Mean Time to Repair
- Cause of failure
- Repair type
- Corrective action taken
- Date of repair

Proper utilization of this information will enable the maintenance/ reliability engineering staff to optimize the preventive maintenance program.

INVENTORY AND PURCHASING

The inventory and purchasing departments need information from the work order system, especially regarding the planned work backlog. If the work is planned properly, inventory and purchasing personnel will know what parts are needed and when they are needed. Good historical information on maintenance material usage will help them establish max/min levels, order points, safety stock, and other settings for maintenance materials. The information required by inventory and purchasing includes:

- Part number
- Part description
- Quantity required
- Date required

Accounting

Accounting needs information from the work order system in order to properly charge the right accounts for the labor and materials used to perform maintenance tasks. The costing system may be different for different locations. However, the following types of accounting information are commonly gathered:

- Cost center
- Accounting number
- Charge account
- Departmental charge number

Upper Management

Upper management is interested in information that can be gathered from multiple work orders. Therefore, this information must be easy to extract from the work order. Summary information should be compiled from completed work orders, work orders in process, and work orders waiting scheduling. If information is not easy to extract, managers might easily spend days gathering the information. Check

Figure 5-2 Work Order Objectives
▶ A method for requesting, assigning, and following up the work
▶ A method of transmitting job instructions
▶ A method for estimating and accumulating maintenance costs
▶ A method for collecting the data necessary for producing management reports

boxes for key information fields can be invaluable for streamlining the process. Computerized systems make this task easier, but only if they are properly designed. In summary, the objectives of the work order system are listed in Figure 5-2.

Types of Work Orders

Any work order system must have several types of work orders. The most common are:

- Planned and scheduled
- Emergency
- Standing or blanket
- Shutdown or outage

Planned and Scheduled Work Orders

Planned and scheduled work orders have already been briefly described. They are the work orders for which a request is made, a planner screens, resources are planned, and the work is scheduled. Work information is then entered in the completion process and the work order is filed. This type of work order, which is the most common, will be discussed in greater detail in Chapter 6, the planning and scheduling chapter.

Standing or Blanket Work Orders

Figure 5-3 lists the purposes of standing or blanket work orders.

These work orders are generally written for 5- to 30-minute quick jobs, such as resetting a circuit breaker or making a quick adjustment. Writing a work order for each of these jobs would bury maintenance in a mountain of detail that could not be compiled effectively into meaningful reports. Standing work orders are written against an equipment charge or an accounting number. Whenever a small job is per-

FIGURE 5-3 STANDING OR BLANKET WORK ORDERS
▶ Repetitive small jobs where the cost of processing the documentation exceeds the cost of performing the work
▶ Fixed or routine assignments where it is unnecessary to write a work order each time it is performed

formed, it is charged to the work order number. The work order itself is not closed out, but remains open for a time period preset by management. It is then closed and posted to history and a new standing work order is opened.

One problem with standing work orders is people sometimes feel as if they are used like credit cards, charging time for the maintenance technicians that is not otherwise accounted for. Occasionally such charges are made, but when the charges are closed out on the work order, offenders can be spotted. Computerized systems make this detection much easier because they can quickly compile a list of all personnel who have charged time to a work order. Some of the more sophisticated systems can even display the percentage of time any maintenance technician charges to a type of work order. If offenders are suspected, it is easy to find them. However, this example is usually the exception, not the rule. Most employees do not abuse a standing work order system.

EMERGENCY WORK ORDERS

Emergency, reactive, or breakdown work orders are generally written after the job is performed. Breakdowns require quick action.

In most cases, there is not enough time to go through the usual planning and scheduling of the work order. The maintenance technician, supervisor, or production supervisor generally makes out the emergency work order after the job is completed. The format of the emergency work order is similar to the work request in that only the brief, necessary information is required. When the work order is posted to the equipment history, it should be marked as an emergency work order, allowing the analysis of the emergency work orders by:

- Equipment ID
- Equipment type
- Department
- Requestor

Analyzing emergency work patterns can help identify certain trends that can be invaluable when planning maintenance activities. The typical flow of a trouble call or emergency work order is pictured in Figure 5-4.

The need for a central call-in point for work requests is to prevent overlapping assignments. If requests are taken at different points, several maintenance technicians may be dispatched to the same job.

When maintenance technicians or supervisors arrive at the job site, they may realize that the job is more involved than the call may have indicated. If the work required is going to exceed a certain cost or time limit, the job is routed back to the planner for analysis. If the work is going to be easier to coordinate and plan by scheduling, the planner takes control of the work order, scheduling it as soon as the material and labor resources are available. This approach allows for cost effective maintenance activities, instead of wasting labor productivity waiting to do a job.

FIGURE 5-4 REACTIVE WORK FLOW
► The trouble call is received by a central dispatch point. ♦ May be the maintenance supervisor
► The call is dispatched to the maintenance supervisor.
► If the repair is going to be over a certain time or cost limit, the supervisor and planner analyze the repair.
► When the logistics are arranged, the repair is performed.

SHUTDOWN OR OUTAGE WORK ORDERS

Shutdown or outage work orders are for work that is going to be performed as a project or during a time when the equipment is shut down for an extended period. These jobs, which are marked as outage or shutdowns, should not appear in the regular craft backlog. This work is still planned, insuring that the maintenance resource requirements are known and ready before the shutdown or outage begins. Such planning prevents delays and maximizes the productivity of all employees. In many cases, the work order information is entered into project management software to run a complete project schedule.

Computerized Maintenance Management Software (CMMS) does not include enough features of project management software to make it an acceptable scheduling alternative. Some vendors have included interfaces to project management systems, which tends to correct this deficiency.

Obstacles to Effective Work Order Systems

Figure 5-5 outlines some of the most common work order problems.

Figure 5-5 Typical Problems for Work Order Systems
► Inadequate or ineffective preventive maintenance programs
► Inadequate labor controls
► Inadequate stores controls
► Poor planning and scheduling disciplines
► Lack of performance measurement
► Inaccurate or inadequate equipment history

Inadequate or Ineffective Preventive Maintenance Programs

These problems can impact the success of a work order system, causing anything from a simple nuisance to total ineffectiveness. Preventive and predictive maintenance programs are keys to operating a work order system. If an organization is in a reactive mode, it has little or no time to operate a work order system. Providing the information necessary to satisfy the work order system takes time. When an organization runs from breakdown to breakdown, it either has little or no time to record the information or any information recorded is generally sketchy and inaccurate.

Companies are in a proactive mode when the work is planned on a regular schedule, with 20% or less emergency activities. This schedule provides supervisors and planners with the time needed to properly utilize the work order reporting system. Without preventive or predictive maintenance programs, it will be impossible to properly utilize a work order information system.

Inadequate Labor Controls

A lack of controls for the maintenance labor resource is a second factor that prevents optimum usage of a maintenance work order system. The following problems are common with labor resources:

- Insufficient personnel of one or all crafts
- Insufficient supervision of personnel
- Inadequate training of personnel
- Lack of accountability for work performed

Without controls in these areas, inadequate or unacceptable resources may be all that is available when the planner tries to schedule the work. Having the labor resources properly controlled is important for a work order system to be effective.

Inadequate Stores Control

A lack of stores controls can reduce the work order system to total ineffectiveness when materials are required. Planners who do not have accurate, timely information concerning the materials in the maintenance stores cannot schedule the work. If the maintenance technician has a work order requiring certain materials, but the materials are not available when they are needed, valuable time is wasted obtaining the materials, thereby lowering the maintenance technician's productivity. Planners, supervisors, and maintenance technicians must have current information about the stock levels of maintenance inventory items. Most consultants believe that maintenance materials are the most essential part of a good maintenance planning program.

Poor Planning and Scheduling Disciplines

Poor planning disciplines affect the work order system because most of the information on the work order is not reliable. In this situation, work orders fall into disuse, resulting in discontinuance of the work order information flow. Job plans must be accurate and realistic if the work order system is to be successful. If companies do not have a work order planning system, they really do not have a work order system.

Lack of Performance Measurement

Lack of performance controls is really a lack of follow up on management's part. Once a job plan or a work order is produced, it should always be audited for compliance. This audit can highlight weaknesses in:

- Planning
- Scheduling
- Supervision
- Maintenance technicians

Any deficiencies can then be corrected. However, if performance controls are not used, the lack of accountability will create a disorganized effort, again allowing the work order to fall into disuse.

Inadequate or Inaccurate Equipment History

Inadequate or inaccurate equipment history hinders the work order system because none of the information used to make management decisions will be reliable. Managers will not be able to base budget projections, equipment repair forecasts, or labor needs on historical standards. The work order system presumably is not being used accurately; equipment history is built from the work order history file. Unless care is taken to see that all posted data is accurate, the unreliable work order system will not be used.

Conclusion

The work order system is the cornerstone for any successful maintenance organization. If work orders are not used, the organization cannot expect much of a return on investment from the maintenance organization. However, work order problems are not all maintenance related. Unless all parts of an organization cooperate and use the system, true maintenance resource optimization will be just a dream.

Chapter 6
Maintenance Planning and Scheduling

A recent survey polled maintenance managers about their top problems. As Figure 6-1 illustrates, over 40% of the respondents indicated that maintenance scheduling was their biggest problem. Maintenance planning and scheduling is one of the most neglected disciplines today. This chapter will explore reasons for this lack of good planning and scheduling as well as some solutions to the problem.

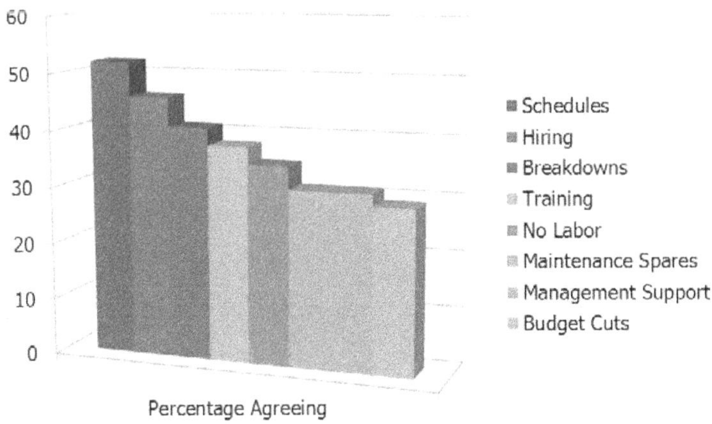

Figure 6-1, Maintenance Problems

Maintenance Planners and Supervisors

One of the major obstacles to maintenance planning and scheduling is management's reluctance to acknowledge that planners are essential to the program. In fact, Figure 1-3 (see Chapter 1) shows that while two-thirds of maintenance organizations in the United States have planners, they fail to use them properly. However, Figure 6-2 shows a hidden problem. Even when organizations have planners, they place responsibility for too many craft technicians on them. Planners should be responsible for 15 (optimum) to 25 (absolute maximum) craft technicians. Supervisors then are responsible for overseeing the work of an average of ten craft technicians. Why the difference between the two groups? This can best be answered by examining the job descriptions for supervisors and planners.

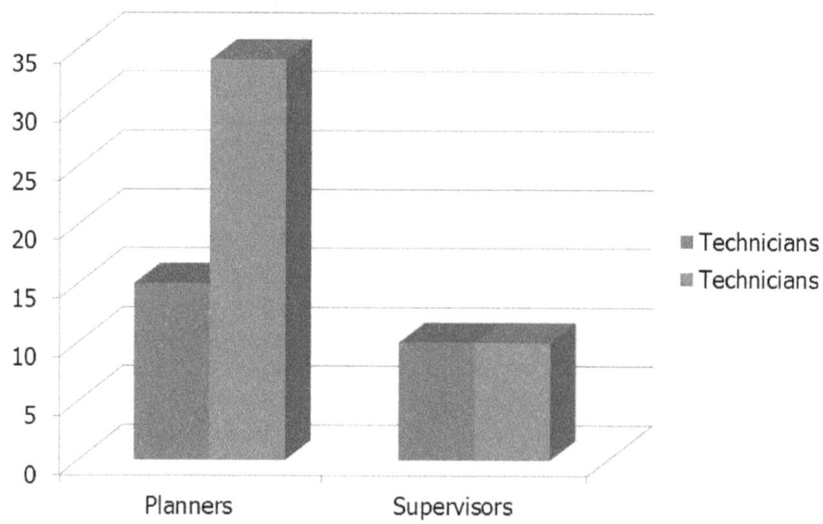

Figure 6-2, Planner/Supervisor Ratios

Job Description for Supervisors

Chapter 3 discussed many of the responsibilities that make up a supervisor's job responsibilities. The following material looks more closely at how supervisors impact work execution.

Among the supervisor's responsibilities is seeing that the maintenance technicians they are assigned to supervise are fully motivated and ready to perform their jobs each day. This does not mean the supervisors must use a command-and-control style, but instead assume the role of a coach. Because the most effective way to motivate is by example, supervisors should be ready to begin at starting time each day and should also be available to assist the maintenance technicians throughout the workday.

When supervisors determine the craft, skill, and crew needed for the job, they are not actually determining what craft manpower is required. This decision has already been made by the planners. Instead, supervisors look at the individual job and match the skill level of the maintenance technicians to the job or, if the job is a larger one, match the right skill level for several crafts and send them all out to the job. Supervisors are responsible for determining who works on each job. The coordination and follow-up of each job requires that supervisors be in the field with the maintenance technicians. They do not sit behind a desk because they cannot see what is going on from there. The safety and quality of the job also require that supervisors be out with the maintenance technicians. Therefore, they must have knowledge of the job skills and techniques, given that quality and safety is their responsibility.

If supervisors are to be responsible for the hiring, firing, and pay reviews of the employees assigned to them, they must be properly trained. Too many times the maintenance managers want to handle the reviews, discipline, commendations, etc. But who best knows the work habits and skills of the employees? It is the direct supervisors, not the next level of management, who know them best. Proper management training will enable supervisors to manage this responsibility effectively.

If supervisors are to recommend improvements and cost reductions, they must be technically competent, understand the jobs their crews are asked to perform, understand the processes they are asked to maintain, and have a basic understanding of the engineering principles involved in the equipment design of the equipment being maintained. Although this appears to be a demand on their technical side, it merely highlights that maintenance supervisors must be technically skilled, if they are to be successful.

Identifying the causes of failures or of repetitive breakdowns highlights the troubleshooting skills of supervisors. If they are not effective in this area, the maintenance technicians on the crew merely become parts changers. Supervisors' feedback to the work order system relating to cause of failure can help make adjustments in the preventive and predictive maintenance programs, as well as give engineering valuable feedback on design flaws. Recommending the necessary skills and training programs for the craft technicians again highlights the interaction between supervisors and their personnel. By knowing what the personnel can and cannot do, supervisors are in a position to recommend what new training is needed to improve their job knowledge and skills. This step improves the attitude of the craft workforce, while enhancing supervisors' ability to manage.

When examining the supervisor's job description, several points become clear. First, supervisors must have in-depth job knowledge of the maintenance skills they are going to supervise. This knowledge is important because many of their tasks require knowing what the maintenance technicians are doing. Second, supervisors must be out on the floor with their assigned maintenance technicians. Unfortunately in the majority of maintenance organizations today, the job of the supervisor has become a "glorified clerical position". This is improper use of the supervisor. Front-line maintenance supervisory positions should be structured so that supervisors spend no more than 25% of their time on paperwork. The other 75% of their time should be spent on the floor with their people. This problem has developed over time. Cutbacks, which in turn eliminate clerical assistance, have forced supervisors to assume duties not in the job description.

Job Description for Planners

An even greater mistake is an organization's hesitancy to dedicate maintenance planners to plan and schedule maintenance activities. The planner's job description, which was discussed in Chapter 3, is clearly different than a supervisor's job. Maintenance departments must be properly staffed if maintenance supervisors are to fulfill their job assignments. To highlight how important it is to have planners, and to show how their jobs differ from maintenance supervisors, it is necessary to review in more detail aspects of the planner's job description

Chapter 3 highlighted the main responsibilities of planners. Now we look more closely at how planners impact the scheduling of maintenance activities. The planner's job starts when a work request is received. Planners review all requests, insuring that they are not currently active work orders. The planners must also clearly understand what the requestor is asking for so that the work plans they develop will produce the desired results. If they are unclear about what is requested, they will visit the job site. This visit serves two purposes. First, it insures the planners will clearly understand what is requested. Second, it gives them time to look for any safety hazards or other potential problems that may need to be documented. If after visiting the job site, they are still unclear about what is being requested, they visit the requestor. This face-to-face discussion will insure that the work is accurately understood before planning begins.

Next, the planners estimate what maintenance craft groups will be needed for the job and also how long they will need. This step is extremely important because these estimates provide the foundation for scheduling accuracy. Planners next decide what materials are needed. Accurate stores information is crucial for this decision. Without reliable information about on-hand quantities, maintenance planning will be inaccurate, resulting in maintenance schedules that are unreliable. Before completing this step, planners insure that all materials are available and in sufficient quantity, before completing this step. They may not find the necessary parts in the storeroom and may need to order them directly from the manufacturer. These parts are referred to as non-stock items. Their delivery date becomes the key to further processing of the work order. Planners insure that all the required resources, including labor, materials, tools, rental equipment, and contractors, are ready before the work is scheduled. This availability eliminates lost productivity because everything is ready before the maintenance technicians begin to work on the job.

Based on prior completions and engineering studies, the planners maintain a file of repetitive jobs. These jobs are performed the same way, using the same labor and materials each time. They are not done on a regularly scheduled basis, but with varying frequencies. The planners build a file, statistically averaging the actual of the labor hours and related costs each time the job is done. This figure becomes the estimate the next time the job is scheduled, thereby increasing the accuracy of the estimates. Planners may also keep the historical file of work orders by equipment. Then when a job comes up that has been done before, they can go to the history file and pull the previous work order. By copying the job steps, materials, and other relevant information from the previous work order, the job planning becomes easier. The planners would look for completion comments to insure the previous job plan did not overlook anything.

Because the planners maintain control of the work order file, they are responsible for developing the craft backlog, a total of all the labor requirements for work that is ready to schedule. This information allows planners to alert management about the need to increase or decrease craft labor. The planners plot the backlog trends for a running six-month chart that allows trends to be tracked. The planners also track the labor capacity for each week so that they can take enough work out of the backlog to prepare the weekly schedule. This total take into account factors such as vacations, sickness, and overtime, and insures that an accurate schedule will be produced.

By matching work from the backlog to the labor availability, planners produce a tentative weekly schedule. They present it to management, who make any needed changes and approve the schedule for the net week. The schedule is then given to the maintenance supervisors at the end of the week, so they can prepare for the next week. Planners do not tell supervisors when they will do each job, or who works on each job; these decisions are the supervisors' responsibility. Planners are responsible for weekly schedules; supervisors for the daily schedules. When the work orders are completed, the planners receive them, note any problems, and file them in the equipment file. The work order file is kept in equipment sequence for easy access to the equipment repair history. Each work order contains the following information:

- Date of repair
- Work order number
- Accumulated downtime

- Cause code
- Priority of work
- Actual labor
- Actual materials
- Total cost
- Year-to-Date costs
- Life-to-Date costs

This information can be complied by management into reports for future decision making. Although it is a cumbersome task annually, computerized systems can compile the information with relative ease. Planners are also responsible for maintaining the equipment information, such as drawings, spare parts listings, and equipment manuals. This information is available to the entire maintenance organization, but is particularly helpful to the planners for their work. Because they have access to the work order files, they can work with engineering to spot any excesses or deficiencies in the preventive and predictive maintenance systems.

A planner has a full-time job, one that is more paperwork oriented than the supervisor's job. Planners should expect to spend 75% of their time on paper and computer work, and only about 25% of their time on the floor, looking over equipment parts or spares. Therefore, no one can be both a planner and a supervisor. Both are full-time jobs, with important and time-consuming responsibilities.

Job Skills for Planners

The job skills a planner should have are listed in Figure 6-3.

First and most important, planners must have good maintenance craft skills. If they are to be effective in planning the job, they must know how to do the job themselves. If job plans are not realistic and

FIGURE 6-3 **MAINTENANCE PLANNER QUALIFICATIONS**
▶ Experienced in craft skills
▶ Experienced in maintenance work controls
▶ Experienced in materials controls
▶ Strong administrative and organizational skills ♦ Sketching skills useful
▶ Shop planners require shop skills
▶ Field planners require multi-field knowledge
▶ Salary grade equivalent to maintenance supervisor

accurate, the program will not be accepted by the maintenance technicians; poor planning increases their work and frustrations.

Planners must also have good communications skills. They are required to interact with multiple levels of management, operations and facilities, and engineering. Poor communication skills can dramatically impact the relationship maintenance has with any or all of these groups. Planners must also have a good aptitude for computer and paperwork because 75% of their time is spent in this type of activity. Some maintenance technicians cannot make this transition. The requirements should be made clear to them before they become planners. In addition, planners must also have the ability to clearly understand instructions. They will be conveying instructions to workers. If they are not clear, how can the planners expect the maintenance technicians to understand them?

Good sketching ability may at first seem superfluous, but planners are often asked exactly what part needs changed. Sketching ability enables them to draw the part quite easily. Thus, sketching becomes an important communication skill and is an indispensable skill for planners to have.

Planners must also be educated about the priorities and management philosophy for the organization. Without a clear understanding, it will be difficult for them to function in a satisfactory manner. However, a good understanding can enhance the entire planning program's performance and contribute overall to the program's acceptance.

Reasons for Planning Program Failures

Figure 6-4 highlights the most common reasons for planning failures.

FIGURE 6-4 REASONS FOR PLANNING FAILURES
► Overlapping responsibilities ♦ Area-Craft-Department ♦ More than one person was responsible and something was overlooked
► The planner is not qualified
► The planner was careless
► The planner did not have sufficient time to plan properly ♦ Trying to plan for too many craft technicians ♦ Involved in reactive work activities

Examining planning failures in more detail can help prevent future programs from meeting with failure. One reason programs fail is weak job descriptions combined with overlapping job responsibilities. This means the first planner thinks the second planner is doing a particular task. Meanwhile, the second planner thinks the first planner is handling it. Thus, no one plans the job and the planning program loses credibility. Eliminating this problem requires strict planning lines. Whether the plans are by craft, crew, department, or supervisors, the responsibilities must be clear, and the plans must be monitored. This problem can easily be eliminated with good management controls in place.

Planners who are not qualified will quickly bring a planning program to its end. Unrealistic or ridiculous job plans will destroy the credibility of the planning program. First, the requirement of having the proper job skills must be met, followed by the rest of the qualifications. Planners should be properly trained and given the opportunity to apply the training. However, for the sake of the entire program, ineffective planner must be removed.

Planners can get careless and job plans will suffer. When this becomes a problem, proper disciplinary procedures should be implemented. These procedures should not be implemented, however, before checking whether management is to blame. Quite often, carelessness by planners can be confused with problems that originate further up the organization.

What is the major reason why planners do not have enough time to properly plan? The ratio of planners to technicians is not correct. As discussed earlier, the ratio should be 1:15 at the optimum. A ratio of 1:20 could possibly be used if the working conditions and type of work planned are optimum. Anything above 1:25 spells certain failure for the program. Consider the steps needed to plan a work order. Could any one person do that for 25 work orders per day? How about 50 work orders per day? Estimate the number of work orders each craft technician completes each day, then multiply by the number of technicians per planner to find the work load for the planner. Overloading the planner is a common problem. Eliminating this problem helps to insure a successful planning program.

Benefits of Planning

Planning offers many benefits. First, it provides cost savings. Table 6-1 lists documented savings that companies have had by switching to planned maintenance from breakdown or emergency maintenance:

Table 6-1 Costs of Planned vs. Unplanned Maintenance		
	Planned Maintenance	Unplanned Maintenance
Job 1	$30,000	$500,000
Job 2	$46,000	$118,000
Job 3	$ 6,000	$ 60,000

In each case, an identical job was performed once in a breakdown mode and subsequently in a planned and scheduled mode. The cost savings from these three examples would pay for a comprehensive planning program for a considerable time period.

In addition to cost savings, planning contributes to an increase in maintenance productivity, which also affects the morale of the workforce. Consider the definition of hands-on time (or wrench time) in Figure 6-5.

The national average for hands-on time is less than 30% for maintenance technicians. In some reactive organizations, it is even below 20%. Why is this happening? Figure 6-6 highlights productivity losses that result from unplanned work.

FIGURE 6-5 **WRENCH TIME OR HANDS-ON TIME**
▶ Wrench Time (or Hands-On Time) is the time the craft technicians physically have their hands on their tools performing the task that they are assigned.
▶ This is the time for which they are actually being paid.

FIGURE 6-6 **LABOR PRODUCTIVITY LOSSES**
▶ Waiting for instructions
▶ Waiting for spare parts
▶ Looking for supervisors
▶ Checking out the work assignment
Multiple trips between the work site and the storeroom to obtain spare parts
▶ Not having the proper tools
▶ Waiting for approval to continue improperly scoped work
▶ Excessive craft technicians assigned to the job

These delays and losses are so common, they need no elaboration. In defense of the maintenance technicians, they are a skilled craft group. Like any other craftsmen, they want to do the best job possible. Lack of cooperation and coordination on management's part impacts their ability to do so. Figure 6-7 lists elements needed to insure the maintenance technicians' ability to take pride in their work activities.

FIGURE 6-7 **PRIDE IN WORKMANSHIP ENABLERS**	
►	Technicians have job instructions.
►	Technicians understand the objectives of the work assignment.
►	The right spare parts are available.
►	The right tools are available.
	Technicians have the clear understanding and support of the operations and facility personnel.
►	They are allowed to complete work assignments when they are started.

All of these elements are part of a good planning and scheduling program. Planning is an integral part of any successful maintenance organization; it affects everything from the bottom line to maintenance technician morale. If you have tried planning unsuccessfully in the past or are planning now, try implementing some of the suggestions from this section. The remainder of this chapter looks at the scheduling component in depth. Remember, however, that without good, accurate job plans, scheduling maintenance work is impossible.

Maintenance Scheduling

In its simplest sense, maintenance scheduling matches the availability of maintenance labor and materials resources to the requests for them from others. If it were that simple, however, maintenance scheduling would not be listed as one of the major problems for maintenance managers. Scheduling involves starting with good job plans; identifying the different stages, or statuses, of the work order; scheduling work when resources are available; and completing the work when scheduled. When planning the work order, the planner needs to track the work order through its various stages, identified by their status codes. Figure 6-8 lists common status codes for work orders.

FIGURE 6-8
WORK ORDER STATUS CODES

► Wait Codes	► Work Codes
♦ Authorization	♦ Ready to Schedule
♦ Planning	♦ In Process
♦ Engineering	♦ Completed
♦ Materials	♦ Cancelled
♦ Shutdown / Outage	

Planners want to insure that a work order has cleared all wait codes before it is given the status "ready to schedule." Scheduling work before it can be started decreases maintenance productivity. Next, planners determine the available labor capacity for the scheduling period. The most accurate formula for determining maintenance labor capacity is shown in Figure 6-9.

FIGURE 6-9
MAINTENANCE LABOR CAPACITY

- Total Gross Capacity (TGC)
  - (Total Technicians X Total Hours Worked) + Overtime Worked + Outside Contract Labor Utilized
- Weekly Deductions (WD)
  - (Weekly Average of Hours Spent on Reactive Work) + (Weekly Average of Absentee Hours) + (Weekly Average of Standing or Routine Work) + (Weekly Average of Miscellaneous Time)
- Net Capacity + TGC – WD

True labor capacity can be compared to a payroll check. It has a gross amount and a net amount. The gross is the hours worked times the pay rate. The net is what is left after taxes, social security, etc. Labor capacity also has a gross amount: total employees times the hours scheduled, plus overtime plus contract workers. However, you can never expect the gross amount of work to be done, any more than you can expect to spend the gross amount of a paycheck. Deductions from the labor capacity include unscheduled emergencies, absenteeism, and allotments for PM work or routine work.

Good scheduling also requires knowing the craft backlog, the amount of work that is waiting for each craft. The craft backlog can be accurately determined only if real world figures are used. Figure 6-10 shows the formula for accurately measuring the craft backlog in weeks.

FIGURE 6-10 MAINTENANCE BACKLOGS
Maintenance staffing levels should be determined by craft backlogs using the following formula.
Craft Backlog $= \dfrac{\text{Open Work Orders Ready to Schedule (total hours)}}{\text{Craft Capacity (weekly)}}$

Accurate backlogs involve open work orders that are ready to schedule, not work that is waiting for some resource before it can be scheduled. Dividing total hours by the craft's net capacity for the week (determined by Figure 6-9) gives the craft backlog in weeks. In turn, knowing the craft backlog in weeks helps to determine the staffing requirements for the craft group. A good backlog is 2-to-4 weeks worth of work. Some companies will allow a 2-to-8 week range. Beyond that level, requestors tend to bump up the priority and circumvent the scheduling process. A craft backlog greater than 4 weeks indicates a need for increased labor, which can be filled by:

- Working overtime
- Increasing contract labor
- Transferring employees
- Hiring employees

A craft backlog of less than 2 weeks indicates a need for reduced labor, which can be met in the following ways:

- Eliminating overtime
- Decreasing contractors
- Transferring employees
- Laying off employees

To properly manage the workforce, it is necessary to trend the backlog over a time period. This step helps to identify developing problems and to evaluate attempted solutions. A good graph should be for a rolling 12-month time period. A sample graph is pictured in Figure 6-11.

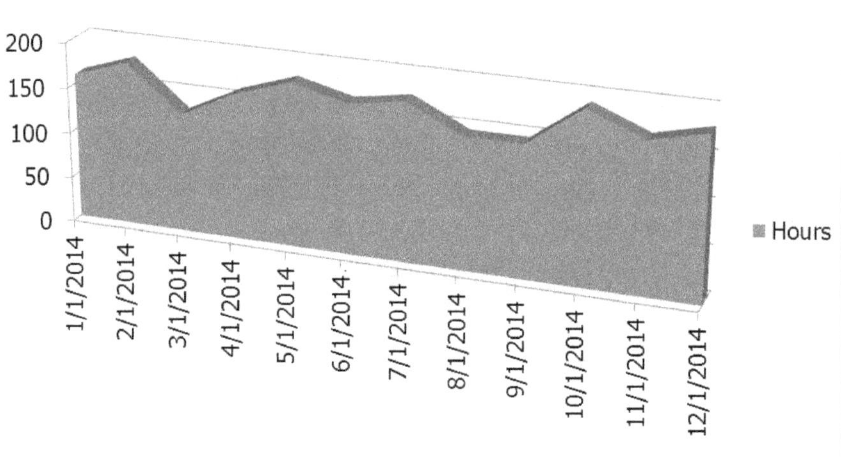

Figure 6-11, Craft Backlog

With this information in hand, planners can now begin preparing the schedule. Remember that planners are concerned primarily with weekly schedules. This time period provides maximum flexibility for handling unexpected delays such as emergency and breakdown work, weather delays, and production rush orders. As they begin the process, planners need to be aware of the considerations listed in Figure 6-12.

FIGURE 6-12 **SCHEDULING CONSIDERATIONS**
▶ Work priority
▶ Work already in progress
▶ Emergency and breakdown work ♦ Historical forecast
▶ Standing and routine work
▶ Preventive maintenance work ♦ Due and overdue
▶ Net craft capacity (weekly)
▶ Size of the craft backlog

Figure 6-12, Scheduling Considerations

Maintenance priorities are based on a variety of factors (see Figure 6-13). This priority measures the criticality or importance of the work.

The simpler that a priority system is, the more widely accepted will be its use. As it grows more complex, fewer people understand it and, in turn, fewer people properly use it. Figure 6-14 shows one of the more complex priority systems available.

This priority system allows for input from both maintenance and production as to the importance of the equipment and the requested work. When the two factors are multiplied together, the final priority is derived. The higher the priority, the faster the work gets done. Some systems even allow for an aging factor, which raises the priority so many points for each week the work order is in the backlog. This prevents Lifers, which are work orders that never get completed.

FIGURE 6-13 BASIC PRIORITY SYSTEM
▶ Emergency or breakdown
▶ Urgent, critical (24–48 hours)
▶ Normal, plan, and schedule
▶ Shutdown, outage, rebuild
▶ Preventive maintenance
▶ Safety
▶ Fabrication

Referring back to Figure 6-13, planners use the status of the work order to begin the listing. All work already in process should be scheduled first in order to eliminate jobs that are already partially completed in the backlog. These jobs would be sorted within this status by priority. The next status would be those work orders previously scheduled, but not started. These also would be sorted by priority within the status. The next would be the work that is ready to schedule, again sorted by priority.

Planners then deduct the hours required to do each work order from the available capacity for the craft group. When they run out of hours of crafts labor available, they can no longer expect the crew to complete any additional jobs for the next week. Those remaining work orders on the list go into the backlog.

The listing might be as follows:

W.O.#	Status	Priority	Date Needed
101	in process	10	
102	in process	9	10/21/03
103	in process	9	10/30/03
104	scheduled	10	
105	scheduled	8	
106	scheduled	5	
107	ready to sched	9	
108	ready to sched	6	
109	ready to sched	3	

FIGURE 6-14
MULTIPLIER WORK PRIORITY AND EQUIPMENT CRITICALITY

Work Priority Code (WPC)	Production Machine Code (PMC)
10. Critical	10. Critical
9.	9.
8.	8.
7.	7.
6. Moderate	6. Moderate
5.	5.
4.	4.
3.	3.
2.	2.
1. Unimportant	1. Unimportant

WPC x PMC = Final Priority

Next, the planners take the schedule to a management meeting where they present what is scheduled for the next week. The maintenance manager, production and facilities managers, and engineering manager may request some changes. Planners make those changes, perhaps deferring some work orders in favor of getting others completed. Once agreement has been reached about the schedule, planners finalize it and distribute copies to all parties involved, usually on the Friday of the preceding week. In this way, they insure that there is complete agreement about the schedule before the week starts.

Conclusion

Figure 6-15 highlights important requirements for maintenance work and maintenance scheduling.

FIGURE 6-15 REQUIREMENTS FOR SCHEDULING
► Good, accurate estimates
► Good work order system, including job instructions, crafts required, and required data
► Accurate craft availability
► Accurate stores information
► Accurate contractor information

The cooperation between the various groups involved will insure that these goals are not just wishes, but will become realities. By following the guidelines here, it should be relatively easy to schedule maintenance work successfully. Figure 6-16 further details the goals that planners should meet for successful scheduling.

FIGURE 6-16 MAINTENANCE SCHEDULING
► Should be 80–90% scheduled
► Should be planned by experienced technicians
► Should be processed as backlog, weekly schedule, then daily work
► Must be flexible enough to accommodate emergency work
► Should not be scheduled until ready

Chapter 7
Preventive Maintenance

If you ask a room of twenty people to write their definition of preventive maintenance, you will get twenty different answers. The term has varied definitions. For our purposes, we define preventive maintenance as any planned maintenance activity that is designed to improve equipment life and avoid any unplanned maintenance activity. In its simplest form, preventive maintenance can be compared to the service schedule for an automobile. Certain tasks must be scheduled at varying frequencies, all designed to keep the automobile from experiencing any unexpected breakdowns. Preventive maintenance for equipment is no different.

The Importance of Preventive Maintenance

The reasons a preventive maintenance program is necessary are listed in Figure 7-1.

Increased automation in industry requires preventive maintenance. The more automated the equipment, the more components that could fail and cause the entire piece of equipment to be taken out of service. Routine services and adjustments can keep the automated equipment in the proper condition to provide uninterrupted service.

FIGURE 7-1 REASONS FOR PREVENTIVE MAINTENANCE
▶ Increased automation
▶ Just-In-Time manufacturing
▶ Lean manufacturing
▶ Business loss due to production delays
▶ Reduction in equipment redundancies
▶ Reduction of insurance inventories
▶ Cell dependencies
▶ Longer equipment life
▶ Minimize energy consumption
▶ Produce a higher quality product
▶ The need for a more planned, organized work environment

Just-In-Time manufacturing (JIT), which is a common strategy in the United States today, requires that the materials being produced into finished goods arrive at each step of the process just in time to be processed. JIT eliminates unwanted and unnecessary inventory. However, JIT also requires high equipment availability. Equipment must be ready to operate when a production demand is made; it cannot break down during the operating cycle. Many companies have advanced from JIT into a "Lean" philosophy, which still has a high demand for predictable availability from the equipment.

Because both JIT and Lean demand the reduction or elimination of buffer inventories (and high costs) traditionally found in U.S. processes, preventive maintenance is necessary to prevent equipment downtime. If equipment does fail during an operational cycle, there will be delays in making the product and delivering it to the customer. In these days of intense competitiveness, delays in delivery can result in lost customers. Preventive maintenance is required so that equipment is reliable enough to develop a production schedule that, in turn, is dependable enough to give a customer firm delivery dates.

In many cases, when equipment is not reliable enough to schedule to capacity, companies will purchase another identical piece of equipment. Then, if the first one breaks down on a critical order, they have a back-up. With the price of equipment today, however, this back-up can be an expensive solution to a common problem. Unexpected equipment failures can be reduced, if not almost eliminated, by a good preventive maintenance program. With equipment availability at its highest possible level, redundant equipment will not be required.

Reducing insurance inventories has an impact on maintenance and operations. Maintenance carries many spare parts in case the equipment breaks down. Operations carry additional spare parts in process inventory for the same reason. Good preventive maintenance programs allow the maintenance departments to know the condition of the equipment and prevent breakdowns. The savings from reducing (in some cases, eliminating) insurance inventories can often finance the entire preventive maintenance program.

In manufacturing and process operations, each production process is dependent on the previous process. In many manufacturing companies, these processes are divided into cells. Each cell is viewed as a separate process or operation. Furthermore, each cell is dependent on the previous cell for the necessary materials to process. An uptime of 97% might be acceptable for a stand-alone cell. But if ten cells, each with a 97% uptime, are tied together to form a manufacturing process, the total uptime for the process is only 71%. Consider:

.97 x.97 = .9409
.9409 x .97 = .9127
.9127 x .97 = .8853

and so on (see Figure 7-2).

This level is unacceptable in any process. Preventive maintenance must be used to raise uptime to even higher levels. Performing needed services on the equipment when required leads to longer equipment life. Returning to an earlier example, an automobile that is serviced at prescribed intervals will deliver a long and useful life. However, if it is neglected — for example, the oil is never changed — it will have a shorter useful life. Because industrial equipment is often even more complex than the newer computerized automobiles, service requirements may be extensive and critical. Preventive maintenance programs allow these requirements to be met, reducing the amount of emergency or breakdown work the maintenance organization is required to perform.

Preventive maintenance reduces the energy consumption for the equipment to its lowest possible level. Well-serviced equipment requires less energy to operate because all bearings, mechanical drives, and shaft alignment receive timely attention. By reducing these drains on the energy used by a piece of equipment, overall energy usage in a plant can amount to a reduction of 5% to as much as 11%.

Quality is another cost reduction that helps justify a good preventive maintenance program. Higher product quality is a direct result of a good preventive maintenance program. Poor, out-of-tolerance equipment never produces a quality product. World class manufac-

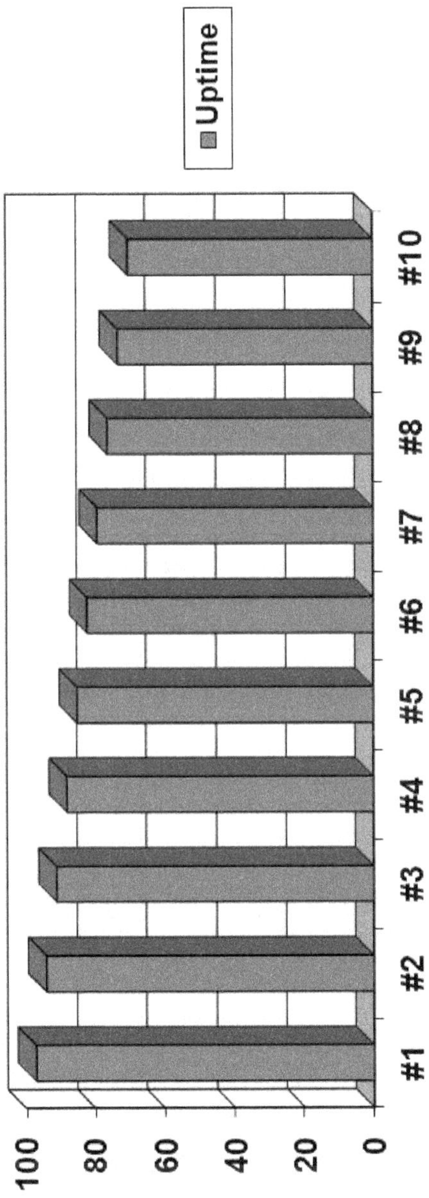

Figure 7-2, Cell Uptime for Ten Machine

turing experts recognize that rigid, disciplined preventive maintenance programs produce high quality products. To achieve the quality required to compete in the world markets today, preventive maintenance programs are required.

If operations or facilities were organized and operated the way the majority of maintenance organizations are, we would never get any products or services when we needed them. An attitude change is necessary to give maintenance the priority it needs. This change also includes management's viewpoint. U.S. management tends to sacrifice long-term planning for short-term returns. This attitude causes problems for maintenance organizations, leading to reactive maintenance with little or no controls. When maintenance is given its due attention, it can become a profit center, producing positive, bottom line improvements to the company.

No preventive maintenance program will be truly successful without strong support from the facility's upper management. Many decisions must be made by plant management to allow time to perform maintenance on the equipment instead of running it wide open. Without upper management's commitment to the program, PM will either never be performed, or it will be performed too little, too late. Thus, management support is the cornerstone for any PM program.

Types of Preventive Maintenance

The various types of preventive maintenance are listed in Figure 7-3.

A good PM program will incorporate all of these types, with the emphasis varying from industry to industry and from facility to facility. This list also provides a progressive step-by-step method for implementing a comprehensive preventive maintenance program.

FIGURE 7-3
TYPES OF PREVENTIVE MAINTENANCE

► The Basics ♦ Inspections, lubrications, proper fastening procedures
► Proactive Replacements
► Scheduled Refurbishing
► Predictive Maintenance
► Condition Based Maintenance
► Reliability Engineering

BASIC PREVENTIVE MAINTENANCE

Basic preventive maintenance — including lubrication, cleaning, and inspections — is the first step in beginning a preventive maintenance program. These service steps take care of small problems before they cause equipment outages. The inspections may reveal deterioration, which can be repaired through the normal planned and scheduled work order system. One problem develops in companies that have this type of program: they stop here, thinking this constitutes a preventive maintenance program. However, it is only a start; a company can do more.

PROACTIVE REPLACEMENTS

Proactive replacements substitute new components for deteriorating or defective components before they can fail. This repair schedule eliminates the high costs related to a breakdown. These components are usually found during the inspection or routine service. One caution: replacement should be only for components in danger of failure. Excessive replacement of components thought but not known to be defective can inflate the cost of the preventive maintenance program. Only components identified as defective or "soon to fail" should be changed.

Scheduled Refurbishings

Scheduled refurbishings are generally found in utility companies, continuous process-type industries, or cyclic facilities, such as colleges or school systems. During the shutdown or outage, all known or suspected defective components are changed out. The equipment or facility is restored to a condition where it should operate relatively trouble free until the next outage. These projects are scheduled using a project management type of software, allowing the company to have a time line for starting and completing the entire project. All resource needs are known in advance, with the entire project being planned.

Predictive Maintenance

Predictive maintenance is a more advanced form of the inspections performed in the first part of this section. Using the technology presently available, inspections can be performed that detail the condition of virtually any component of a piece of equipment. Some of the technologies include:

- Vibration analysis
- Spectrographic oil analysis
- Infrared scanning
- Shock pulse method

The main differentiation between preventive and predictive maintenance is that preventive maintenance is more of a basic task, whereas predictive maintenance uses some form of a technology.

Condition Based Maintenance

Condition based maintenance takes predictive maintenance one step further, by performing the inspections in a real-time mode. Sensors installed on the equipment provide signals that are fed into the computer system, whether it is a process control system or a building automation system. The computer then monitors and trends the information, allowing maintenance to be scheduled when it is needed. This eliminates error on the part of the technicians, who would otherwise make the readings out in the field. The trending is useful for scheduling the repairs at times when production is not using the equipment.

Reliability Engineering

Reliability engineering, the final step in preventive maintenance, involves engineering. If problems with equipment failures still persist after using the aforementioned tools and techniques, engineering should begin a study of the total maintenance plan to see if anything is being neglected or overlooked. If not, a design engineering study should be undertaken to study possible modifications to the equipment to correct the problem. Incorporating all of the above techniques into a comprehensive preventive maintenance program will enable a plant or facility to optimize the resources dedicated to the PM program. Neglecting any of the above areas can result in a PM program that is not cost effective.

The Benefits of Preventive Maintenance

The most prominent benefit of a PM program is the elimination of costs related to a breakdown that occurs during production or operation. The cost to repair the equipment is easy to calculate,

(Number of Workers X the Hours to Repair) + Material Costs

However, this cost is not the total picture. Total costs for a breakdown or an unexpected outage of the equipment are listed in Figure 7-4.

FIGURE 7-4 BREAKDOWN MAINTENANCE COSTS
► Operator Time Loss ♦ Time to report the failure ♦ Time for maintenance to arrive ♦ Time for maintenance to make the repairs ♦ Time to restart the equipment
► Maintenance Costs ♦ Time to get to the equipment ♦ Time to repair the equipment ♦ Time to get back to the dispatch area
► Cost of repairing or replacing the filed part or component
► Lost production or sales cost or both
► Cost of scrap due to the maintenance action
► If a facility, the cost of lost productivity of the occupants of the building

Most of the points in this figure are self-explanatory, but the costs may still be difficult to calculate. One of the largest intangibles is the price paid for late or poor quality deliveries made to a customer. Lost business or unhappy customers can have a dramatic impact on future business. Published reports indicate that customers lost because of poor quality can average as high as 10% of sales per year. When this figure is coupled with averages as high as 30% of the manufacturing budget going for quality problems and rework, it becomes easy to stress the importance that a PM program can have on quality.

Financial justification for PM programs is not difficult. The costs and savings are listed in Figure 7-5.

FIGURE 7-5 **PREVENTIVE MAINTENANCE**
► Initially increases ♦ Maintenance labor costs ♦ Replacement parts costs
► Eventually decreases ♦ Scrap/Quality costs ♦ Downtime / Lost production costs ♦ Lost sales costs

Types of Equipment Failure

What type of equipment failure is it best to address with a PM program? There are at least four different types of failures: infant mortality, random failures, abuse, and normal wearout.

Infant Mortality

Infant mortality is a failure occurring in the first few hours of component life. This failure is understood by the electronics industry where burn-in of components is common. In this case, the failure occurs when initial voltage is applied to a circuit, but the component is not up to standard. It is impossible to design a PM program to prevent this type of failure.

Random Failures

Random failures occur without notice or warning. This type of failure, which is difficult to predict, is engineering or materials related. Because of their unpredictability, a PM program cannot be designed to prevent them.

Abuse Failures

Abuse or misuse failures generally result from a training or attitude problem. No preventive maintenance program can prevent this type of failure.

Normal Wear Out

Normal wear out is the type of failure that preventive maintenance programs can be designed to prolong or prevent. These failures occur progressively over a relatively long period of time. PM programs can be designed to spot signs of wear and take appropriate measures to correct the situation. Normal wear is allowed to progress, either a) there is no real consequence of a failure or b) a component is replaced just before normal wear causes a failure.

Developing Preventive Maintenance Programs

Once the decision has been made to develop a PM program and strong management support has been obtained, the steps listed in Figure 7-6 should be followed.

FIGURE 7-6 STEPS FOR STARTING PREVENTIVE MAINTENANCE PROGRAMS
▶ Determine the critical equipment units or systems.
▶ Identify the components that make up the units or systems. ♦ Belts, chains, gear drives, electrical systems, electronic systems, fluid power systems,etc.
▶ Determine the preventive maintenance procedure for each type of component.
▶ Develop a *detailed* job plan for each of the procedures.
▶ Determine the schedule for each of the preventive maintenance procedures.

DEVELOP THE CRITICAL EQUIPMENT UNITS AND SYSTEMS

The first step of designing a PM program is to determine the critical units and systems in the plant that will be included in the PM program. Maintenance managers know that having the PM program cover every item in the plant or facility is not cost effective. There are certain components, not part of critical processes, which are cheaper to let run to failure than to spend money maintaining. Critical items should be identified and cataloged for inclusion in the PM program.

Identify Components

After the units have been identified, it is necessary to break the equipment down to the component level. It is then easier to develop standard PM procedures. For example, most V-belt drives use the same maintenance procedures. Identifying all the V-belt drives makes it easier to write the procedures and apply it to all other V-belt drives, making any small changes required to customize the procedure for each piece of equipment. In determining the procedures for each type of component, several sources of information may be consulted. Figure 7-7 lists some of these sources.

Figure 7-7 **Source Information for PM Procedures**
► Manufacturers ♦ Will provide maintenance, lubrication, and overhaul schedules
► Review Work Order Histories ♦ Can provide information on breakdown frequencies and repairhistori es
► Detailed consultation ♦ Operators, supervisors, craft technicians can provide equipment-specificinformationanddetails

Depending on the quality of the information, historical records can provide the most accurate information because it comes directly from the specific plant environment. A word of caution is required if manufacturer information is used: the manufacturer's recommended tasks and recommended frequencies tend to call for too much maintenance performed too often. Most manufacturers would like to see their equipment serviced as often as possible. However, excessive servicing inflates the cost of the PM, making it too expensive in some cases. Manufacturer information is good to use for guidelines, but should not be considered hard, fast rules.

Develop Detailed Procedures and Job Plans

The next step after deciding on the tasks to be performed is to develop detailed procedures on how to perform each task. These procedures should include the following information:

- The required craft
- The amount of time required for the craft
- A listing of all materials required
- Detailed job instructions, including safety directions
- Any shutdown or downtime requirements

Figure 7-8 Preventive Maintenance Task Estimates
► Preparation Time ♦ Lock Out / Tag Out
► Travel Time ♦ To and from the site
► Restrictions ♦ Confined space, fire watch, etc.
► Actual Performance Time
► Area Clean Up Time

All of this information should be detailed. For example, when estimating the time required to perform the PM, the estimate should consider all of the factors listed in Figure 7-8.

Determine the Schedule

The more detailed the information in the plan for the PM, the easier it is to schedule. PM schedules are generally integrated into the overall maintenance schedule, unless there are personnel dedicated only to performing the PMs. In either case, more accurate estimates and material requirements lead to more accurate schedules and, in turn, more successful PM programs. Inaccuracies lead to overscheduling, resulting at first in missed PM or altered frequencies and, ultimately, in additional breakdowns or a failure. This unsuccessful PM program loses management support, spelling failure for the program.

Types of PM Tasks

Once a PM program is developed, a final decision still needs to be made regarding the types of PMs to be performed. Figure 7-9 lists sample types of PMs.

Figure 7-9 Types of PM Tasks
► Mandatory or Non-Mandatory ♦ Regulatory or non-regulatory
► Pyramiding or Non-Pyramiding ♦ Fixed date or based on last due date
► Inspections or Task Oriented ♦ Just inspections or change parts

MANDATORY AND NON-MANDATORY TASKS

Mandatory PMs are ones that must be performed at all costs when they are due. They may involve OSHA, safety, EPA, and license inspections, among others. Non-mandatory PMs are inspections or service PMs that can be postponed for a short time period or even eliminated for the present cycle without resulting in immediate failure or performance penalty. Each PM task should be designated in one of these categories.

PYRAMIDING AND NON-PYRAMIDING TASKS

Pyramiding PMs are generated each time they come due. When there is already a PM due and the next one comes due, the first one should be canceled, with a note written in the equipment history that the PM was skipped. The new PM should have a due date from the canceled PM written in, so that it is understood how overdue the task is.

Some companies, however, choose to make their PMs floating or non-pyramiding. They follow the same scenario as described above, except there is no notification that the PM was missed. The previously uncompleted PM is thrown away and the new one (without any carry over information) is issued and placed on the schedule.

This approach results in an incomplete picture of the status of the PM program. Instead of showing incomplete or late PMs, the system shows everything being up to date. Increased failures will result. The PM program, showing no apparent effect on equipment operation, will be looked at as worthless, and lose credibility with management. Eventually the PM program is likely to be canceled or reduced in the PM program. To prevent this, some form of a tracking mechanism must be provided to validate the frequencies of the PM.

INSPECTIONS OR TASK ORIENTED

Inspections will involve only filling out a check sheet and then writing work orders to cover any problems discovered during the inspection. Task-oriented PMs allow the individual performing the PM to take time to make minor repairs or adjustments, eliminating the need to write some of the work orders when turning in the inspection sheet. For scheduling purposes, a time limit should be set on how long each task should take. Some companies set a time limit of one hour of work in addition to what the PM was planned for. If the job is to take any longer, the individual should return and write a work order for someone to perform the repairs. This approach prevents the PM program from accumulating labor costs that should be attributed to normal repairs.

Preventive Maintenance Program Indicators

Figure 7-10 lifts indicators that show when a PM program is ineffective.

Each of these indicators can be used as an argument for improving an existing program or to justify starting a program. For example, if equipment utilization is below 90%, the equipment is not being serviced correctly. If a PM program is in place, it needs rapid adjustment before management decides it is of no value and does away with it. High wait time for the machine operators when the equipment fails indicates a major failure. A good PM program should detect major failures before they occur. If there are numerous major failures, the PM program must be changed to address the problems before support is lost for the program. If breakdowns can be traced to lack of lubrication or adjustments, the PM program is again to blame. The program

Figure 7-10 Indicators of PM Problems
► Low MTBF ♦ Equipment breaks down frequently
► High MTTR ♦ Usually catastrophic breakdowns when they occur
► Maintenance related quality problems ♦ High scrap and rejects
► Failures relegated to neglect of maintenance basics ♦ Poor results from lubrication, inspection, or fastening procedures
► Regulatory issues related to maintenance ♦ OSHA, EPA, FDA, ISO-9000
► General equipment condition poor ♦ Decrease in capital equipment life

should be quickly adjusted to address the problems. A good PM program should remedy all lubrication and service-related failures.

Summary

The most common reasons for PM programs being discontinued or being ineffective are listed in Figure 7-11. All of these problems have been discussed in this text.

The set of bulleted items listed under "Priority" in Figure 7-11 relate to the PM programs not being given enough priority within the maintenance organization. When a PM program is implemented, all must commit to giving it the dedicated effort needed to make it work. The bulleted items listed under start up issues indicate the PM program had an improper start. Corrections in these areas should be made as soon as a deficiency is discovered in order to keep the management support necessary to make the PM program successful.

The items listed under "Static Program" in Figure 7-11 indicates the need to keep the program flexible, allowing for changes in the equipment during its life cycle. As the equipment ages, its requirements vary. The PM program must adapt to reflect these needed changes if it is to be cost effective. Failure to change or adjust the program will result in a PM program that may be successful for a few years and then begin to be more expensive than necessary. Too much or too little maintenance, along with their related failures, will be noticed. The program develops problems with management support. All PM programs must be closely monitored if they are to continue to be successful.

FIGURE 7-11
PM PROGRAM FAILURES

- Priority
  - Used as busy work or fill-in jobs
  - Schedule conflicts
  - Failure to stay in schedule compliance
- Start Up Issues
  - Wrong equipment selection, poorplanning,insufficienttaskdetails
- Static Program
  - Failure to make life-cycle adjustments or perform failure analysis

Preventive maintenance is the foundation for a successful maintenance program at any Best Practice company. If sufficient effort is placed in the preventive maintenance program, then many other practices can be improved at a minimum cost and effort.

Chapter 8
Maintenance Inventory and Purchasing

The inventory and purchasing functions have a greater impact on maintenance productivity than any other support group. How do inventory and purchasing affect the maintenance organization? Figure 8-1 on the following page lists ways that poor inventory control can affect maintenance productivity.

Chapter 6 noted that maintenance work should be planned. Part of the job plan for maintenance is detailing all the materials required to perform the work, insuring they are in stock and available before the work is scheduled. The list in Figure 8-1 includes common delays in finding or transporting spare parts. If the job is properly planned, these delays will be eliminated.

Information for Inventory Planning

What does maintenance need from inventory and purchasing in order to be effective planning the work? Figure 8-2 lists the minimum information required.

FIGURE 8-1 MATERIAL-RELATED DELAYS
▶ Waiting for materials
▶ Travel time to obtain materials
▶ Time required to transport materials
▶ Time required to identify materials
▶ Time to find substitute materials
▶ Time required to find materials in area stores locations
▶ Time required to process a purchase order
▶ Lost time due to: ♦ Other crafts delayed due to materials ♦ Wrong materials planned and delivered ♦ Wrong materials ordered ♦ Materials out of stock

REAL-TIME PARTS INFORMATION

On-line or real-time parts information is necessary to plan maintenance activities. When selecting parts for a job, planners must know whether they are in stock, out of stock, in transit, etc. Planners must have current information. If the work is planned based on information that is days, weeks, or months old, then craft technicians could experience all of the delays listed in Figure 8-1 when they go to pick up the parts. If the information the planners have is current, then they will know what action can be taken. For example, the minimum parts information planners need includes:

- Part number
- Part description
- Quantity on-hand

- Location of part
- Quantity reserved for other work
- Quantity on order
- Substitute part number

Although planners could use other information as well, the list above will cover the majority of the jobs. However, if this information is inaccurate or unreliable, the planners will have to physically check the store each time work is planned. This time-consuming activity will lengthen the time needed to plan a job properly; eventually the planners will not be able to plan all of the work required.

Hard Copy Catalogs

A current hard copy listing of all the parts carried in the stores for maintenance must be provided, even if the inventory system is com-

Figure 8-2 Maintenance Requirements from an Inventory System
▶ Real time parts information
▶ Hard copy of the inventory catalog
▶ Equipment "Where Used" listing
▶ Parts usage ♦ By cost center, department, equipment, etc.
▶ Accurate on-hand quantities
▶ Projected delivery dates

puterized. This is particularly important when the storeroom is not staffed on 2nd or 3rd shift, or over weekends. This catalog allows all maintenance personnel access to the stores information, specifically the line item catalog. It is not used for planning because the on-hand or order information would be dated. But the line item catalog allows maintenance personnel to know if parts are stock items or non-stock items. Knowing this distinction can help expedite the process if a certain part is needed. It prevents the delay of having the maintenance personnel looking through the storeroom to find a part that is not stocked. This situation occurs frequently, especially during a breakdown or emergency repair. Providing maintenance stores catalogs at key locations can help eliminate costly delays.

"Bill of Material " Listings

Equipment Bill of Material (BOM) listings provide information, by equipment, of all the spare parts carried in the stores. This listing is important in several ways. First, it allows planners quick access to the parts information during the planning process. They will always know on what piece of equipment the work is being performed. They can then look up the part information quickly. Not finding the part on the list points out a possible need of adding it to the list of spare parts, and requesting that stores now carry it in stock. Second, this list is important during a breakdown or emergency situation. When a part is needed, a scan of the spare parts list can save time looking for the part.

Accurate On-Hand Quantities

Planners must have accurate on-hand information. If the inventory system indicates that a store has sufficient supplies of a part to do a job, the planners may send craft technicians to get them. When the technicians discover the parts are not there, the inventory system loses credibility. This loss will impact the inventory system's usefulness to maintenance. If the planners or technicians have to go to the store location personally each time a part is planned or requested, the maintenance department will experience a tremendous loss of productivity.

PROJECTED DELIVERY DATES

Projected delivery dates are also important. No store will always have every part when it is requested. Knowing when the part will be delivered allows planners to schedule the work based on that date. Therefore, the delivery performance of the vendors must also be reliable. If a job is scheduled for a certain week, but parts are not delivered as promised, there will be another loss of maintenance productivity.

Inventory System Requirements

Figure 8-3 lists other requirements that improve the quality of the inventory system so that it can enhance the productivity of the maintenance department.

TRACKING ASSETS

Returns in a production inventory system indicate how many items have been sent back to the vendor for some reason. In a maintenance inventory system, returns indicate how many parts have been planned for a job and issued to a work order, but were not needed and, therefore, were returned to the stores for credit. This indicator becomes a measure of the planner's performance. If a planner always orders too many parts for each job, then the inventory stock level will

FIGURE 8-3 MAINTENANCE INVENTORY SYSTEM REQUIREMENTS
▶ Tracks balances for all items, including issues, reserves, and returns
▶ Maintains a parts listing for equipment
▶ Tracks item repair costs and movement history
▶ Cross reference spares to substitutes
▶ Can reserve items for a job
▶ Can notify a requestor when items are received for a job
▶ Can generate a work order to fabricate or repair an item
▶ Can notify when item reorder is needed and track the reorder to receipt
▶ Can track requisitions, purchase orders, and special order receipts
▶ Can produce performance reports such as inventory accuracy, turnover, and stockouts

be higher than required. This level ties up capital in spares that the company could put to better use elsewhere in the business.

In many companies, tracking the movement of rebuildable spares is important. The third listing in Figure 8-3 shows that this information should be tracked through the stores information system. This information is used both for accounting purposes and for the repair history that is important to making repair and replace decisions. Tracking movement through inventory is the easiest way to maintain this information. Furthermore, stores personnel maintain the parts in storage.

It is easier to let them control the spares, provided that maintenance can get access to the information when necessary.

Notifying When Parts Are Received

The sixth listing in Figure 8-3 is also important to planners. In many cases, planners order a part for a job, holding the job till the parts come in. Because they may plan twenty or more work orders per day, after several weeks, they may have dozens of work orders waiting on parts. It is important to have a method of notifying them when the parts are received and which work orders they were reserved for. This seemingly small detail can literally save planners hours of work.

Monitoring Performance

The last item in Figure 8-3 is also important. As with any other part of the organization, stores and purchasing should be monitored for performance. The indicators mentioned in this item are useful for tracking performance levels. Poor performance by stores and purchasing will have a dramatic impact on the maintenance organization. Maintenance managers should receive copies of any inventory and purchasing reports. The maintenance manager then can refer to this information for comparison with maintenance performance. Any conflicts between the two groups can then be discussed and remedied.

Organizing Maintenance Stores

Maintenance stores locations are critical to the productivity of the maintenance personnel.

Types of Stores Options

Figure 8-4 lists the two types of maintenance stores options.

Figure 8-4
Options for Maintenance Stores

- ► Centralized Stores
  - ♦ Reduced record keeping
  - ♦ Reduced stores labor costs
  - ♦ Increased maintenance travel
    - ▪ Lost productivity
- ► Area or Decentralized Stores
  - ♦ Reduced maintenance travel
    - ▪ Increased productivity
  - ♦ Increased record keeping
  - ♦ Increased stores labor costs
  - ♦ Increased inventory levels

These options are similar to the maintenance organizational structures. In fact, most companies that have area maintenance organizations will have area stores locations. Maintenance productivity is then increased by eliminating travel time to get spare parts. However, maintenance stores do not need to be located at each maintenance area shop. Stores may be located between several maintenance areas and still have acceptable travel time to get spare parts. Centralized stores are good for central maintenance organizations. There should be no unnecessary delays in the maintenance technicians obtaining their spare parts. If a central stores location is used, it should be staffed correctly in order to avoid creating delays for people trying to get material out of the stores.

Types of Maintenance Spares

Maintenance has many different types of spares that need to be tracked through the inventory function. Some of the most common categories are listed in Figure 8-5.

Figure 8-5 Types of Maintenance Spares
▶ Bin Stock — Free Issue
▶ Bin Stock — Controlled Issue
▶ Critical or Insurance Spares
▶ Rebuildable Spares
▶ Consumables
▶ Tools and Equipment
▶ Residual or Surplus Parts
▶ Scrap or Useless Spares

Using these categories will help maintenance managers insure that correct controls are placed on the more important items, whereas those with less importance have fewer controls.

Bin Stock Items

Bin stock items are materials that have little individual value, but high volume usage. Examples include small bolts, nuts, washers, and cotter pins. These items are usually placed in an open issue area. Their usage is not tracked to individual work orders as are larger items.

The best way to maintain the free issue items is the two-bin method. Items are kept in an open carousel bin where the craft technicians can get what they need when they need it. When the bin becomes empty, the store clerk puts the new box in the bin, and orders two more boxes. By the time the bin is emptied, the boxes are delivered and the cycle starts over again.

Bin stock controlled issue items are similar to the free issue items, except their access is limited. The stores clerk will hand the items out, although still not requiring a requisition or work order number for the item. The stock levels should be maintained similarly to the free issue stock levels using the two-box method.

Critical or Insurance Spares

Critical or insurance spares are those items that may not have much usage, but due to order, manufacture, and delivery times must be kept in stock in case they are needed. An important factor in the inventory decision is the cost of lost production or amount of downtime that will be incurred if the part is not stocked. If this cost is high, stocking the item will be better than risking the cost of a breakdown. Because the cost of these items is usually high on a per unit base, they must receive proper care while in storage. This means a heated, dry, weather-proof storage area. If the spare remains in storage for six months, a year, or even longer, good storage conditions will prevent its deterioration.

Rebuildable Spares

Rebuildable spares include items like pumps, motors, gear cases, and other items for which the repair cost (materials and labor) is less than the cost to rebuild it. Depending on the size of the organization, the spare may be repaired by maintenance technicians, departmental shop personnel, or sent outside the company to a repair shop. These items are also generally high dollar spares and must be kept in good environmental conditions. Their usage, similar to the critical spares, must be closely monitored and tracked. Lost spares of this type can result in considerable financial loss.

Consumables

Consumables are items that are taken from the stores and used up or thrown away after a time period. These items include flashlight batteries, soap, oils, and greases. Their usage is tracked and charged to a work order number or accounting code. Historical records may be studied and charted to determine the correct levels of stock to carry for each item. If problems develop with the stock level, the inventory level can be adjusted on a periodic basis.

Tools and Equipment

In some companies, tools and equipment are kept in the stores location or in a tool crib, then issued like inventory items. With a tool crib, the tools are returned when the job is finished. The tool tracking system monitors the tool's location, who has it, what job it is being used for, and the date the tool is returned. This type of system is used only to track tools with a specifically high value or when only a relative few tools are available in the entire company. This system should not be used to track ordinary hand tools.

Residual or Surplus Parts

When maintenance is involved in construction or outside contractors are doing construction work in the plant, surplus or residual materials are generally left over. Because there is no place else to put them, they end up in the maintenance stores. These residual or surplus items can become a problem in the stores. If the parts are not going to be used again in the short term (one to six months), they should be returned to the vendor for credit. If they are going to be used, or are critical spares, they should be assigned a stock number and stored properly. A word of caution: Storing these items just to have them is expensive. Store rooms can easily become junkyards, costing their company money that most employees do not realize. We will examine these costs later in this chapter.

Scrap or Useless Spares

Over a period of time, all stores accumulate scrap and other useless spare parts. At least once per year the stocking policies should be reviewed. If there are scrap items, get rid of them. One method is to go through the store with a supervisor, a planner, and a craft technician, identifying every recognizable item. Then pile the remaining items outside the storeroom with a sign "If you recognize any of these parts, put an identification tag on them." After two weeks, simply scrap anything left over.

Classifying Spares

Whether managers realize it or not, keeping spares is costly. Another method of classifying spares is the A-B-C analysis. This system is outlined in Figure 8-6.

"A" items are high dollar, insurance-type items that must be kept in stock. These items should have strict inventory policies regulating their use and movement. Because there are relatively few "A" items, controlling them is not difficult. "B" items are more numerous than "A" items, but not as costly. These items also should be controlled in a strict tracking method. Controlling the "A" and "B" items covers

FIGURE 8-6
A-B-C ANALYSIS

- "A" items are
  - 20% of the stock items
  - 80% of the total inventory value
- "B" items are
  - 30% of the stock items
  - 15% of the total inventory value
- "C" items are
  - 50% of the stock items
  - 5% of the total inventory value

about 50% of the total inventory items; but about 95% of the inventory costs. The "C" items are the open bin issue items. They make up about 50% of the total number of items, but only about 5% of the cost. It is a waste of time and energy to try to control the "C" items at the same level as the "A" and "B" items. The monetary return will not justify the labor necessary to process the paperwork.

One additional note on maintenance storerooms: Some companies hold the philosophy that all maintenance stores should be open. This philosophy is incorrect. Accurate and timely inventory information must be kept; controls must be placed on movement of certain maintenance spares. An open store, with no monitoring of the individuals having access to the stores, eliminates any controls. Parts can be used without anyone knowing where they went. Someone may move them within the stores and no one else will know where they are. This type of system is expensive and will not allow a maintenance organization to effectively use its materials. A closed store, at least for "A" and "B" items, is critical to successfully improving maintenance stores.

Inventory Costs

The cost of maintenance inventories has already been discussed. Figure 8-7 lists some additional costs, the hidden costs of inventory.

FIGURE 8-7 HIDDEN INVENTORY COSTS
▶ Cost of capital tied up in the spare part: 15%
▶ Cost of operating the warehouse space, property tax, energy cost, insurance, maintenance fees: 5%
▶ Cost of space occupancy, rent, and depreciation: 8%
▶ Cost of inventory shrinkage, obsolescence, damage, theft, and contamination: 5–10%
▶ Inventory tax: 1–2%
▶ Cost of labor to move in and out: 5–10%
▶ Total real cost of carrying an inventory item per year: 30–40% of the item's value

The total cost of carrying an item in stores may be as high as 30–40% of the value of the item per year. A company with an inventory of $10 million spends $3 to $4 million each year to maintain that inventory — a staggering amount. Figure 8-7 summarizes the critical reason why it is so important companies should carry only as much of each item as is really required. Anything over that amount is waste that is deducted directly from the corporate bottom line. Inventory control is extremely important and should not be overlooked in any effort to improve a maintenance organization.

COST SAVINGS

Because the costs of inventories are so high, what other efforts can be made to curb or control these costs? Figure 8-8 highlights several areas where savings have been realized by many companies.

FIGURE 8-8 COST SAVINGS CONSIDERATIONS
▶ Standardizing plant equipment
▶ Standardizing suppliers
▶ Consignment arrangements
▶ Locating stores at key areas
▶ Specifying maintenance supplies and suppliers
▶ Reducing or eliminating obsolete spare parts
▶ Reducing the amount of spoilage
▶ Controlling "vanished" spare parts

Standardization of equipment, supplies, and suppliers have proven to be large sources of savings for organizations. For example, standardizing equipment can help reduce inventory. Suppose a plant has fifteen presses, each made by a different manufacturer. Few, if any, of the parts would be interchangeable. Fifteen sets of spares would be needed, one set for each of the presses. Imagine the total cost for the inventory for such an arrangement. If, however, the fifteen presses were from the same manufacturer, many parts would be interchangeable. Instead of fifteen sets of spares, perhaps only five sets would be needed because the likelihood of more than five of the presses needing the same part at the same time would be very small. Consider the savings in just the carrying costs for the other ten sets of spares, then multiply this number times the number of different types of multiple equipment items in the plant. Savings can quickly add up to a very large amount.

Meanwhile, studies indicate that by consolidating supplies and suppliers, companies can save large percentages of their total inventory costs. This is one area where we can learn from the Best Practice companies. They keep the number of suppliers low and receive better prices and service. The suppliers receive more business. The simplification of these relationships helps all involved. Yet this type of savings is virtually an untouched area in many industries. Reduction of obsolete, spoiled, or vanished parts is accomplished through better inventory controls and closed storerooms.

These points cannot be overemphasized. A large savings can be made from inventory controls. At the same time, these controls improve the service that maintenance receives from the inventory and purchasing function.

Maintenance Controls

Unfortunately there are many organizations in which maintenance and stores/purchasing do not cooperate. A recent survey showed that only 50% of the organizations polled allowed maintenance any controls over its inventory. This rate is alarming because maintenance is responsible for budgeting for repair materials. Thus, It is being held responsible for something it cannot control. Figure 8-9 highlights minimum controls that maintenance should have over its own inventories.

FIGURE 8-9 AREAS THAT MAINTENANCE SHOULD CONTROL OR INFLUENCE
▶ Item issue quantity
▶ Return policy for unused materials
▶ Storage for rebuilt materials
▶ Order points and quantities
▶ What components to stock and where to stock them

If maintenance cannot have these controls, it should not be held responsible for controlling any costs, because it will not be able to do so. Unfortunately, many organizations are controlled by internal politics. Maintenance usually loses in this type of environment. Inventory and purchasing often influence upper management to a point that

FIGURE 8-10 ORGANIZATIONAL ORDER
▶ The maintenance function serves The company owners and shareholders.
▶ The inventory and purchasing functions Provide a service to the maintenance function.

negates the effectiveness of the maintenance organization. Figure 8-10 highlights an important principle of organizational control.

When maintenance is pushed aside or overlooked, the entire organization suffers. If, however, the organization places its emphasis in the right areas, allowing maintenance to control its own resources, then maintenance can become a profit center, enhancing corporate profitability.

Chapter 9
Management Reporting and Analysis

It has been said that:

- To manage, you must have controls.
- To have controls, you must have measurement.
- To have measurement, you must have information.
- To have information, you must collect data.

Data...information...facts. Whatever term you use, you need knowledge to make good decisions.

Computerized Maintenance Management Systems (CMMS) and Enterprise Asset Management (EAM) Systems

In the maintenance / reliability community, the terms CMMS and EAM are used interchangeably. For a point of reference, a CMMS tends to be a departmental system and an EAM system tends to be more of a corporate system. For the purpose of this text, we will use the CMMS acronym, except where the material is specific to an EAM system.

The goal of a computerized maintenance management system (CMMS) is to produce quality data to help maintenance/ reliability professionals make good decisions.

CMMS Modules

Even as a company implements a CMMS, data collection begins. Consider the various modules that make up a comprehensive CMMS:

- Equipment
- Inventory
- Purchasing
- Personnel
- Preventive maintenance
- Predictive maintenance
- Work orders
- Contractors
- Rebuilds

The data for reporting in a CMMS is derived from accurate data input into all of the CMMS modules.

Figure 9-1 shows the basic relationship of these modules to one another. The relationship of some of the modules will be discussed in the following section.

Equipment. To use this module properly, you must identify each piece of equipment—or facility location—that you want to track for costs and repairs. For example, you need the financial information stored in the equipment history when making repair/replacement and other cost decisions. Data provided by the other modules can determine the accuracy of the financial information.

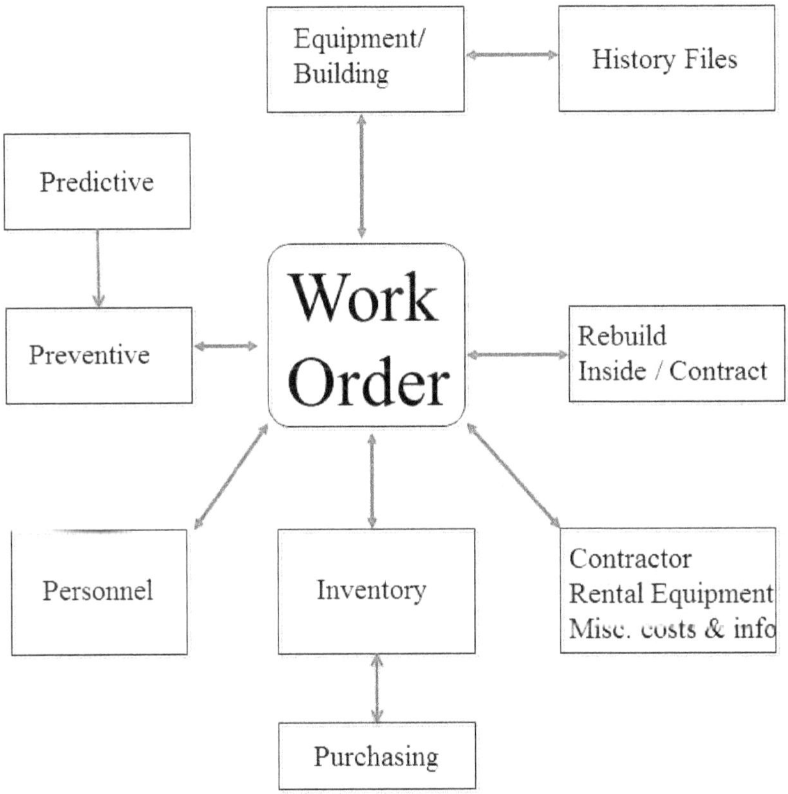

Figure 9-1, CMMS Basic Modules

Inventory. To use this module, you must identify the spare parts carried in each store at your facility. The data you need includes, but is not limited to:

- Part number
- Part description (short and extended)
- On-hand, reserved, on-order, max-min, etc.
- Locations
- Part-costing information
- Historical use

Information from the inventory module ensures that your CMMS will contain accurate material-costing information for each piece of equipment or facility location.

Purchasing. This module is associated with the inventory module. It provides planners with a window into the ordering information. This module includes the following information:

- Part number
- Part description
- Part-costing information
- Delivery information, including the date
- Related vendor information
- The ability to order non-stock materials

The importance of this module becomes clear when you begin to plan a job and do not know when you will receive the part. Furthermore, this module is important for estimating job costs without knowing new part costs.

Personnel. This component allows a company to track specific information about each employee. The required data includes:

- Employee number
- Employee name and personal information
- Pay rate
- Job skills
- Training history
- Safety history

The data from the personnel module insures that the facility will post accurate labor costs to work orders and equipment history.

Preventive Maintenance (PM). This module allows you to track all PM-specific costs. The costing information comes from the personnel and inventory databases. Important information stored in this module includes:

- Type of PM (lubrication, calibration, testing, etc.)
- Frequency required
- Estimated labor costs (from the personnel module)
- Estimated parts costs (from the inventory module)
- Detailed task description.

The collection of this data insures accurate service information and costing each time a worker performs a PM task. A CMMS can also project labor at material resource requirements for calendar-based PM tasks.

Work Orders. With this installment, a user can initiate different types of work orders, or track the work through completion. This module also allows you to allot costing and repair information to the correct piece of equipment or facility location. The use of the work order requires information from all other modules of the CMMS. Information required on a work order includes:

- Identifying the equipment or facility location where the work is being performed
- Identifying the labor requirements (personnel)
- Identifying the parts requirements (inventory)
- The priority of the work
- The date the work must be finished
- Contractor information
- Detailed instructions

In order to be effective, the work order requires information from all the modules. Without accurate information, the work order cannot collect the required data. Therefore, the work order cannot post accurate information to the equipment history. In turn, without accurate data in the equipment history, the maintenance manager cannot make timely and cost effective decisions. All other reports, which are derived from the data stored in the various CMMS modules, also require accurate information.

The success of a CMMS depends on the timeliness and accuracy of collected data and the use of that data by the company. If information is inaccurate or used incorrectly, the system fails.

Time Frame for Reports

Once a company purchases a CMMS, how long does it take before accurate and informative reports can be produced? The answer depends on how long the company needs to develop accurate data. In a survey conducted by Engineer's Digest, 52% of the respondents indicated that 1 to 11 months were needed to make their system fully operational. (Detailed results of this survey were published in a special insert in the April 1992 issue of Engineer's Digest.) ReliabilityWeb (www.reliabilityweb.com) has conducted a more recent survey (in 2008) with almost the exact results. More specifically, the survey showed that 40% of respondents took more than one year to make their system fully operational (see Figure 9-2).

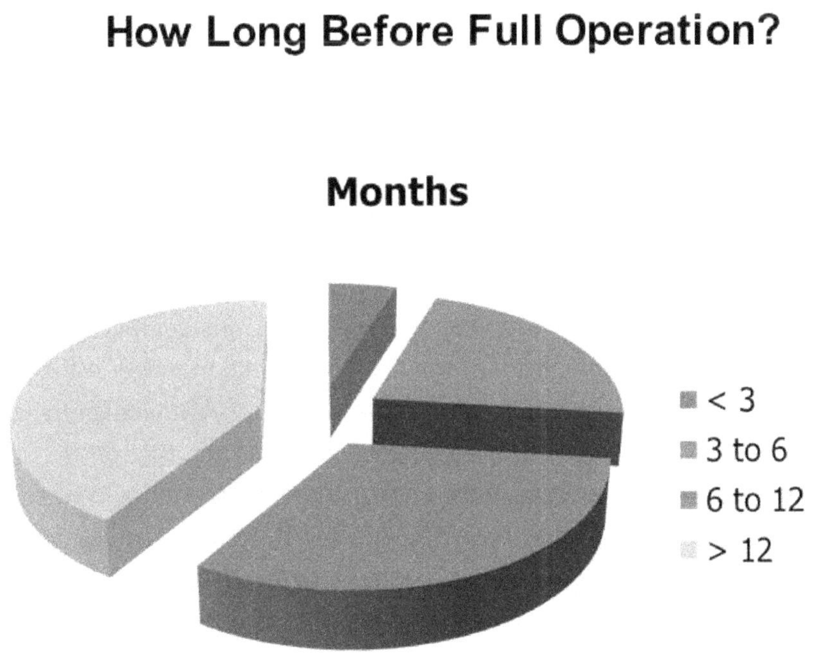

Figure 9-2, How Long Before Full Operation?

Even before a facility implements a CMMS, the information collected will still have some value. But until the system is fully utilized, the data will not be accurate. For example, if only certain departments are on a CMMS (a typical pilot implementation problem), then the data from these departments may be accurate. However, in areas where a crossover or combination with another area or craft exists, the data may be incomplete or distorted.

A CMMS should provide a completely integrated data collection system. However, even many mature users are not getting complete—and thus accurate—data from their CMMS. In the Engineer's Digest survey, 921 respondents were asked about the use of the inventory, purchasing, and personnel modules of their systems. The majority of the respondents indicated they were using less than 70 percent of their systems. The ReliabilityWeb survey showed a similar trend with only 1/3 of the survey respondents effectively using their CMMS for inventory and purchasing functions.

When companies use corporate systems, such as the EAM system, the data may not be posted accurately in the equipment history. In fact, in most cases, the data are inaccurate or not posted. Thus, the equipment history is incomplete or inaccurate. Consider this example. When you take your car in for repairs, the service manager estimates the time and cost of the job (work-order planning). You accept the estimate, and the service shop begins the work. When the job is complete, you receive a shop order with a complete breakdown of each part used and its related cost. The bill (work order) also shows the number of hours the mechanic worked and the hourly rate. The total equals labor and parts. You expect this bill each time you go to the garage for any work. If your bill showed only the final price with no breakdown, you would not accept it. Apply that concept to your EAM System. Are your EAM system reports giving you accurate cost breakdowns for your equipment?

Here's another example. When using a CMMS, if you do not supply planners with closely integrated inventory information, then they cannot be sure the stores' information is accurate. This is especially true if the information is updated only once a day or once a week. This situation arises many times when other corporate systems are interfaced to a CMMS. Workers can waste time looking for a part that is supposed to be in the stores, when, in fact, another worker used that part the previous day or shift. This delay may seem inconsequential. However, when downtime can cost $1,000 or even $100,000/hour, these types of delays may mean the difference between profit and loss for the entire company.

When it comes time to consider replacing your car, do you look only at the labor charges you have made against it for its life? Do you look only at the parts used? No, you consider the whole picture—labor, materials, its present condition, etc. These same principles must carry over in the asset-management programs at companies. However, companies have set the CMMS information flow so the material or labor costs are not shown on the work order or equipment history. Therefore, decisions are based on inaccurate or incomplete data, and such decisions will include mistakes. The financial implications of these decisions can spell disaster for a company. They can force that company into a condition where it cannot compete against other companies that make full use of a CMMS and, thus, obtain the subsequent cost benefits.

The Solution

If a company is collecting data incorrectly, it needs to re-evaluate the system, determining also if the data is incomplete or missing. In addition, it needs to determine what parts of the CMMS are being used incorrectly or not at all.

By evaluating the answers and then working to provide accurate data collection, a company's CMMS use will benefit its bottom line. In today's competitive marketplace, it is unacceptable to make guesses when data is available. The cost benefits obtained making correct decisions will help make a company more competitive. Wrong decisions can put a company out of business by taking away a competitive position.

What CMMS reports should a company use? Some systems are available with no reports, while others have hundreds of "canned" reports. The deciding factor is to use reports that are needed to manage the maintenance function. If the report does not support or verify an indicator used to manage maintenance, it is not beneficial. Reports produce hundreds of pages of data that can overload the maintenance manager. If maintenance is measured by the estimated vs. actual budget, but the CMMS cannot produce a budget report, then the system is not supporting the organization. With CMMS reports, too many can be just as bad as too few.

Because management requires measurement and measurement requires data, each company must use its CMMS fully to obtain data. Without data, it is just someone's opinion. Discussions require factual data. Arguments occur when emotions and opinions enter the discussion, not factual data. Whether employees at a company have discussions or arguments may mean the difference between being a world-class competitor or a second-rate company.

Preceding chapters have shown where data is collected in a maintenance organization. The work order is the key document for collecting maintenance information. But having the information is not important. It must be in a usable form. The work order information should be used to produce reports, providing management with the information necessary to control and manage the maintenance function. The information should be concise and specific. Broad lists of information can be too time consuming for a manager to study. Analysis and exception reports using the information are vital to the management of maintenance. Beyond just maintenance's needs, there are the needs of the entire company. Inventory, purchasing, engineering, and plant management all need from maintenance.

Reports are time consuming to compile manually. In organizations with any appreciably-sized staff, a computer will be necessary to compile them. The reports may be output from a simple database or spreadsheet, sorted and compiled. Companies using a CMMS have an advantage because many of these reports are included in the system. If they are not, a report writer can be used to construct these reports. Computerized reports have another advantage, especially with the relational databases; they can produce meaningful graphic representations of the information. These graphs can be more helpful in describing trends and patterns than columns of figures. They can also be included in reports to upper management. As many reports as possible should be graphed.

Maintenance Reporting

The following maintenance reports are organized into five groups. The first three groups are reports that need to be reviewed by the maintenance staff daily, weekly, or monthly. The fourth group is made up of general information reports and the fifth group by reports that should be produced on an "as needed" basis.

Daily Reports

This group of reports should be produced daily for review by the appropriate maintenance personnel and managers.

Work Summary Report

This report lists work orders currently in progress as well as work orders that have been closed out in the last day. It provides a quick look at the prior day's activity. The report should show estimated vs. actual totals for the following categories:

- Maintenance labor
- Maintenance materials
- Equipment downtime

The report should be divided into two sections: work orders completed and work orders still in process. Each section should be sorted by priority of the work: emergency, planned and scheduled, preventive maintenance, and routine. This summary of work performed allows managers to quickly answer any questions that operations may ask during a daily review meeting.

Schedule Progress Report

This report lists only the work scheduled for the week with the present status of each work order. Like the work summary report, it should show actual vs. estimate figures. The difference is that only work orders appearing on the schedule will be listed in this report. This report should conclude with a summary showing the number of work orders scheduled to be closed out during the week versus those that have been closed out. A percentage could even be included as percent of scheduled work completed. This report allows the manager to monitor the progress of the schedule completion and make any adjustments necessary during the week to insure schedule completion.

Emergency Report

The emergency report lists all emergency or reactive work requests in the last day. It should be a two-part report. The first should be a line summary showing labor hours and materials dollars used. The second part should show which craft technicians worked on the job, what parts were used, and any other detail or completion notes. This report gives managers a quick look at the breakdowns and emergency work for the last day. If a particular job needs clarification, they can reference the second part of the report to get the required details.

Reorder Report

This report lists all inventory items that have reached their reorder point during the last day. Depending on the maintenance and purchasing relationship, this list is used to generate purchase requisitions or purchase orders. If maintenance is not involved in the ordering, the report may be only an information report for maintenance. If the organization is multi-warehouse, this report should be divided by warehouse, allowing the option of transferring between warehouses instead of placing new orders. The report should show on-hand, reserved, and minimum quantities for each item listed.

Meter-Based/Predictive Work Order Report

This report lists all meter-based, predictive, and condition based preventive maintenance work orders generated during the prior day. Depending on the sophistication of the PM system in use, these may be manually generated or automatically generated through real-time interfaces to computer systems. In either case, these work orders need quick attention, usually within a week. Some may even require breaking in to the present schedule to avoid a serious failure. This report should list the equipment, type of PM, type of work to be performed, and labor and material requirements.

Personnel Summary Report

The personnel summary report lists all employees who worked during the previous day, the hours they worked on each work order, and any overtime that was worked during the period. The report gives supervisors a quick review of the previous day's activity. The purpose is to insure that each employee worked and was credited for the proper number of hours worked the prior day.

Work Order Listing

This report lists work orders currently in the backlog. It should be available in several different ways, for example by requesting department, by equipment ID, by craft, by supervisor, and by planner Each report should be sorted in descending order from work orders in process to work orders just requested, allowing for complete information on all work when required.

Weekly Reports

In addition to the daily reports, each week additional reports are required to properly manage maintenance. This section provide examples of these reports.

Schedule Compliance Report

This report compares results from the previous week's activity to the schedule for the same (current) week. It should begin with all work completed, all work not completed, and finally all work not started. This format allows for detailed analysis of the week's activity. The report should also compare time allotted for emergency activities versus the time actually used for emergency activities. This comparison indicate why more or less work was completed than scheduled. The report also compares work scheduled to be worked on versus total work completed. This summary figure can be used as the efficiency percentage. Tracking this percentage over a 6- to 12-month time period provides a complete picture of scheduling efficiency.

PM Compliance Report

This report lists the PMs scheduled to be completed for the previous week and their present status. It may be broken down by:

- PMs completed as scheduled
- PMs started as scheduled, but not completed
- PMs scheduled, but not started

The report gives a quick overview of the status of the preventive maintenance program for the previous week. A detailed section of the report may also show who performed the work, what parts were used, completion comments, any related work orders that were written because of the PM, etc.

PM Due/Overdue Report

This report lists all overdue PMs. It should sort the PMs from oldest to newest. If possible, the report should start with a specified time period, for example, PMs that are over eight weeks overdue, and then sort the remaining PMs down to the ones that just became overdue last week. This structure allows for a quick look first at numbers that are critically overdue, and eventually at ones just overdue. The report is most beneficial when a summary line of the numbers and percentage of work contained in each overdue category is listed. The report can be a good indicator of the condition of the PM program.

Schedule Projection Reports

These reports are a related series that should be produced upon setting the next week's schedule. With the work orders identified for the schedule, the information for these reports should be pulled from the work order estimates.

Required Manpower. This report should show the required craft hours for the work scheduled. The report should first list, by craft, a summary of total hours required. The second part of the report should be the detailed description of the work. The report gives supervisors a quick or detailed overview of their craft groups for the next week. If the schedule is produced by crew or department, then the report should be listed the same way.

Required Downtime. This report lists all equipment downtime required for the work scheduled for next week. It should be divided by equipment, department, and line or plant, depending on the layout of the plant or facility. The report is extremely important, especially for JIT, MRP, MRPII, and CIM installations. Inputting this information into the production control system begins the process of integrating maintenance management.

Required Parts. This report lists all parts required for the work that is scheduled for next week. It should be given to the stores/inventory personnel to insure that parts are ready. If the company practices staging, this report is good to use to pick the materials and send to the staging area.

Required Contractors. This report lists all contractors required for the next week's schedule. It should match the contractors with the work orders they are required to complete. This format reduces the report's size when individual contracts have multiple work orders.

Required Equipment. This report lists all rental, lease, or specialized in-house equipment required to complete the work on next week's schedule. It allows planners or coordinators to insure that the equipment is either ready for delivery or delivered before the work is started.

Required Tools. Companies that have tool rooms for storing general maintenance tools generate this report, which lists the required tools for the week and the work orders for which they will be needed. This information enables the tools room to insure that all the tools required will be in good repair and ready for use.

Work Order Status Report

This report lists all work orders in the backlog. It should be available in the several formats listed below along with a brief explanation of each:

By Department. This report is distributed to the departmental supervisors so they can see the status of all work requested for their department. Thus, they can see the status of their "pet" projects without continually calling maintenance.

By Equipment. This report allows quick access to all work presently requested for all equipment. As a quick reference report for anyone requesting work, it helps prevent duplicate work orders in the system by allowing recipients to review if a similar job is requested for the equipment.

By Planner/Supervisor. This report indicates the workload for both planners and supervisors. Balance can be achieved in the organization by offloading work from one planner or supervisor to others during peak work times.

Past Due Work Order Report

This report lists all work orders in the current backlog sorted by the date needed, ideally from the most overdue to the least overdue. Work orders without this information should not be included. All status work orders, except those completed, should be included. If, for example, a work order is being held up due to materials unavailable, planners may be able to find some way of expediting or substituting the materials to complete the work order. The report can also provide good feedback information to engineering, stores, and operations.

Backlog By Craft/Crew/Department

This report tracks the amount of work for each craft or crew group. In some organizations, the work may even be backlogged by department. It should be a summary type report, showing the total hours for each craft group and then the number of weeks' work. It could also be a trending report, showing the weekly backlog for each craft group for the last year. In this way, it can highlight trends and seasonal peaks, helping management make good, justifiable staffing decisions.

Monthly Reports

In addition to daily and weekly information, other information becomes more meaningful looked at on a monthly basis. Some of these reports are described in this section.

Completed Work Orders Report

This report is a two-part report of all work orders completed during the month. The summary portion shows total work orders closed, actuals versus estimates for man-hours, labor costs, material costs, contractor costs. This information should be available in several ways: sorted by requesting department, by equipment type, by equipment ID. The summary will help in spotting trouble areas. The detail portion is a work order by work order listing, sorted in the same way as the summary portion. This information helps managers investigate any discrepancies in detail.

Planning Effectiveness Report

This report highlights the effectiveness of the planners. It shows all the work orders for each planner, comparing actual costs with planned costs. In some computerized systems, it may also specify a tolerance percentage, reducing the amount of information that managers would have to review. A summary by planner is beneficial, concluding with a listing of the most effective to least effective planner. Any major discrepancies should be reviewed at a detailed level.

Supervisory Effectiveness Report

This report is identical to the planner's report except it focuses on the effectiveness of the supervisors. It should show all the work orders for each supervisor, comparing actual costs with planned costs. In some CMMS, it may be helpful to specify a tolerance percentage, reducing the amount of information that managers would have to review. A summary by supervisor is beneficial, concluding with a listing of the most effective to least effective supervisor. Any major discrepancies should be reviewed at a detailed level.

Downtime Report

This report compares actual downtime with estimated downtime for all closed work orders for the month. The report is important because it includes not only the planned and scheduled work, but also the emergency and breakdown work. Rather than measuring planning and supervising effectiveness, it measures the effectiveness of the entire organization. PM programs, planning, supervising, stores, purchasing, and organizational coordination all play a part in controlling downtime. A specified tolerance could help keep the report shorter. Any major discrepancies should be investigated, with appropriate correctional measures taken.

Budget Variance Report

This report compares the actual figures for all maintenance expenses to the budgeted figures for each category. Depending on the plant or facility budgeting procedures, they may be broken into categories such as labor, material, contractor, and tools. The report provides an opportunity for managers to correct potential problems before it is too late.

Craft Usage Summary Report (By Department and Type of Work)

This is one of the most meaningful reports that can be produced for a analyzing a maintenance organization. It first summarizes the work that was performed in each department (line, building, cost center, etc.) by craft. The summary should include the total resource expenditures. The second part should list percentages of the total resources expended for the entire plant or facility and what percentage was used by each department. The final part should show what percentage of the work done in each department was emergency work, planned work, PM work, or other, as tracked by maintenance managers.

General Information Reports

Some reports list general information and are kept for reference. Although by themselves they do not provide analysis capability, they form the catalog for each category of information.

Equipment Listing

This report lists all equipment that maintenance is responsible to maintain. It has two formats — summary and detailed. The summary is a one-line-per-item list of equipment, including categories such as ID, name, type, and location. The detailed format lists one item per page (or more pages, depending on the level of detail kept). Each page should include all of the information kept on the equipment. This report can be anything from a file folder to a computerized database.

Cross Reference Listing

This report lists all parts stocked in inventory as spares. It may be a listing by equipment of all spares used for an equipment item or by part number, the latter which then lists all the equipment on which the part is used. Both formats have value. The equipment listing helps planners know what parts are carried as spares whereas the parts/equipment listing helps inventory find items that may otherwise be of questionable value to carry as spare parts.

Parts Master List

This report lists all spare parts carried in inventory. It may be produced in two formats: by part number or by part description. Both formats are necessary because there will be times the part number is known and times when it is not. The descriptive lookup helps craft technicians or other personnel who are not familiar with the part number. Each listing includes most of the information about the spare part and may require a page or more.

Employee/Craft/Crew Listing

This report lists all the information about the maintenance personnel. It may be produced by employee ID number or on a last name/first name basis. Each listing then has all the personnel information, which may require a page or more per employee.

Analysis/Decision Justification Reports

This last series of reports are prepared for management on an as-needed basis. They are not simple lists, but instead use advance statistical calculations to produce intelligent output. These reports do not make decisions, but the information, combined with good management skills and insight, can help manage maintenance successfully.

Statistics Report

This report is useful for spotting trends in breakdowns. It lists, either by equipment ID or by type of equipment, the downtime, what caused the downtime, and what effect it had on operations. For example, a report could look at all air conditioning units to see what the most common cause of failure was for a certain time period. Another report may focus on plant lathes to consider the most common cause of breakdowns that led to operational delays of ten hours or more. Careful analysis of this report can lead to improvements in preventive maintenance programs, and even in purchasing policies.

Resource Requirements Forecast (Open)

This report forecasts the known requirements for maintenance activities for a specified period. If the specified time period exceeds the work backlog, it basically becomes a PM forecast. The report lists the work in the backlog, from the oldest to the newest requests. (PMs that are forecast to come due in the specified time period will be included.) Labor, materials, contractor, and other costs will be listed for the forecast period. This information can help in short range (4 to 8 weeks) financial planning. It can also help related groups if they know what portion of their resources will be required from maintenance in the short term.

EQUIPMENT HISTORY REPORT

This report allows the detailing of any work performed on any specified equipment. The report should offer users selection criteria, so they do not have to go through all the work orders to find the desired information. Some categories for sorting the information include:

- Type of work (PM, emergency, planned, etc.)
- Equipment component
- Type of failure
- Type of activity (adjustment, calibration, part change, etc.)
- Date range

This report can help identify specific problems and the number of occurrences for any given equipment item.

REPAIR LENGTH REPORT

This report identifies any repair — whether emergency, P.M., or routine — that has a downtime greater than the amount specified at the start of the report. This information enables users to look at just the repairs that had a measurable effect on operations or facilities. The report should be able to be sorted by equipment, type of equipment, department, craft, or any other field that is important to a particular organization.

Repetitive Repairs Report

This report analyzes repairs that have been performed on the same equipment component when the repairs have been conducted more than the number of times the user enters. The report should list the number of times the repair occurred, how long the equipment was down each time, and what caused the equipment to fail each time. The report is useful for analyzing repetitive failures for any equipment component. If the report is expanded, it can show the failure rate for similar components on various equipment items. This information may be useful for supporting purchasing or outage activities.

MTBF/MTTR Report

This report calculates the Mean Time Between Failures (MTBF) and the Mean Time To Repair (MTTR) for a piece of equipment. The report should be able to break the information down to the component level. This information is useful for setting frequencies for PM programs and planning production schedules. The report can produce the MTBF and MTTR information for failure causes and equipment components. These sections help to highlight any problem component or consistent failure problem. Because trouble areas will be clearly highlighted, the report also helps plan the use of maintenance resources.

Breakdown Analysis Report

This report analyzes all breakdown and emergency work requests. It can be prepared for a specific equipment item or for all the equipment plant wide. It should include, and be able to be sorted by, the following specification fields: equipment type, line, shop, building, account code, failure, cause, repair, and solution The report enables maintenance managers to find specific problem areas and work to correct them. By knowing the largest problems first, resource usage can be optimized.

Top Ten Report

The previous report highlights specific problems on equipment, this report highlights the top ten equipment items in equipment downtime, labor costs, material costs, or contractor use. It should be available by plant, department, building, line, account code, or other appropriate field. This report highlights the specific problem equipment, whereas the previous report can then be used to further analyze the problem.

PM Route Listing

In some plants and facilities, PM tasks are arranged in a route for optimizing travel time on the part of the maintenance technicians. This report lists a particular route and all of the PM tasks associated with it. By analyzing this route with others in the same areas, managers may be able to make adjustments or changes that optimize the technician's travel time.

Overdue PM Report

This report shows all overdue PMs currently outstanding for any equipment item. The report can be sorted by fields such as individual equipment ID, department, line, building, and account code. It should used to highlight problems with missed or overdue PMs. Coupled with the breakdown analysis report, it can spot breakdown problems that are related to poor PM follow-up.

Outage/Shutdown Report

This report allows users to specify the shutdown or outage; it then lists all of the work orders planned for that outage. Users then have all the work orders, the requirements for the work orders, and the resources necessary to perform the work. The information can be fed into the project management program for detailed scheduling with PERT or GANTT charts.

PM Efficiency and Compliance Report

This report has two parts. The first part compares the planned versus the actual for completed PMs. It can be sorted by equipment, department, crew/craft, building, or individual equipment item. The comparisons should include the following areas: planned vs. actual hours, planned vs. actual materials, and planned vs. actual downtime. Users should be able to specify a percentage deviation, allowing for a specific report to highlight problem areas. The second part of this report is the compliance report for the PM frequency. By specifying a percentage deviation allowed, the user should then get a report selected by the same parameters as the first part of the report. This report will then list the scheduled frequency for the PM versus the actual frequency. This information helps the manager in examining compliance and scheduling accuracy for the PM program.

Parts Activity

The parts activity report shows the inventory activity for either a specific warehouse or for all warehouses. It can list activity for a specific part, range of parts, or all parts. The report helps managers analyze the activity of a given location in order to insure proper staffing of inventory and warehouse distribution personnel.

Slow Moving Parts

This report is used to find those spare parts that are carried in inventory, but that are not having sufficient activity to maintain in stock. Managers would not include critical spares in this analysis; by definition, they should be slow moving. But others not having activity in a specified time period could be marked for reduction in on-hand quantities or elimination from inventory. This report can be organized by warehouse or for all warehouses as well as for part groups or types of parts.

Issue/Returns Report

The issues/returns report shows the parts issued to a work order compared to the items returned to the stores as not needed for the work order. The report should also list any materials issued to the work order that were not planned. The report becomes a measure of the effectiveness of the planner's ability to estimate the parts required for a work order. The report should be sorted by planner. The columns for each planner can then be totaled and an effectiveness percentage calculated. This report will benefit both inventory managers and the maintenance manager.

Inventory Over Max Report

Almost all inventory items have maximum/minimum levels set for them. This report lists any stock items whose on-hand quantities are higher than the pre-set maximum quantity. It will highlight any overages. Then appropriate measures can be taken to reduce these quantities.

EOQ Report

The economic order quantity is the optimum quantity to be ordered when purchasing an item. It is beyond the scope and purpose of this text to go into an explanation of the formulas used to calculate this number. The EOQ is an effective method for ordering maintenance spares. This report examines the inventory and lists the EOQ for the inventory items. The information from this report can be then used to adjust this quantity in the part record.

Stock Out Report

This report will sort in descending order the inventory items that have had stock outs during the specified time period for the report. Items having multiple stock-outs give an indication of a problem. These should be cross-referenced to the activity and planning reports in order to find and correct the problem. Stock outs can have high downtime-related costs and should be corrected as quickly as possible.

Inventory Optimization Report

Because there are cost trade-offs in inventory, this report looks at the entire picture and recommends stock levels. It is similar to the EOQ report except it highlights the financial penalties associated with the decision. For example, it uses carrying costs, desired service level, downtime costs, lead times, and other categories to calculate the differing costs of maintaining stock levels. It should list several options for the user, highlighting the costs for more or fewer items in stock.

Reorder List

This report lists all inventory items that are in need of re-order. It is often used where maintenance does not have a computerized tie-in to the purchasing department. It will list by warehouse all parts below their minimum quantity. The report can then be reviewed and forward to purchasing for action.

Parts Receipt List

This report lists all parts that have arrived and been processed by receiving. The quantity received, where they have been stored, and the date should all appear on the report.

A-B-C Classification Report

Chapter 8 on inventory described the A-B-C classification system. This report categorizes the inventory items into the appropriate classification. Combined with several of the activity reports, it can help managers adjusts inventory levels.

Price Change Report

This report lists all items that were ordered at one price, but received at a different one. This difference indicates either a change in price or a mistake by the vendor. The purchasing buyer can then take appropriate action to determine the cause of the price change. Once the matter has been resolved, the inventory records can be updated with the correct price.

Parts Warranty Report

This report lists the warranty of each part in the inventory. This information can be useful if high activity is shown on a part with a lengthy warranty. Investigation may show the justification for a claim against the manufacturer for a refund. Close monitoring of the warranties can result in sizable refunds in some cases.

Restocking Report

This report is used in a multiple warehouse plant or facility. It shows items in stock at the main warehouse that need to be distributed to satellite stores locations. It lists the parts and where they need to go. After this report is run, the reorder report also needs to be run because some items may now fall below their minimum.

List Parts Issues

This report is similar to the parts activity listing, except it lists what work order goes with each part. It is often listed by work order number. More often, it is listed by part, with each work order the part was issued to also listed.

Parts Adjustment Transaction

This report is a listing made after a physical count inventory has been taken. It shows the difference between what was counted and what the inventory record indicates was on-hand. In some computerized systems, this report is used to accept the physical count; it then lists the cost variance for each item and finally the report.

Overdue P.R. Report

This report lists the purchase requests that are older than a time period specified by the requestor of the report. This report insures that buyers consolidate the purchase requests into a purchase order in a timely manner. Because many maintenance activities are dependent on quick turnaround times, this is a key performance measurement report.

Overdue P.O. Report

This report shows all purchase order that are past the expected delivery date. This report is important in that many maintenance activities are based on these dates. When purchase orders are overdue, the buyer must call the vendor, check the status of the order, and do what is necessary to expedite the order.

Vendor Performance Analysis

This report analyzes the on-time delivery, price, and total amount of business conducted with each vendor. This information can be used to work for discounts, insist on quality, or other performance related issues. The report helps determine vendor of choice for purchasing critical items.

P.O. History Report

This report is for tracking the receipt history of a purchase order. Because many purchase orders contain multiple line items, the parts are received in partial shipments. This report lists each purchase order and all shipments received to date against it. It also serves as an indicator of the vendor performance.

Outstanding P.O. Report

This report lists all outstanding purchase orders for any or all vendors. These may or may not be overdue purchase orders. The report also complements the vendor performance report, which has historical information whereas this report provides the current outstanding purchase orders.

Cost Variance Report

This report helps to synchronize all of the part prices throughout the inventory/purchasing/receiving process. Any parts with a difference in price in any of these areas should appear on this report. This information helps insure accurate parts costing in the maintenance work order system.

Overtime Report

This is a two-part report, the first part being a listing of employees with the total overtime they have worked for the specified reporting period. The second part of the report is a call list; it is based on company policy for contacting employees to work overtime. These reports can be time savers for supervisors.

Canceled PM Work

This report is a selective report organized by equipment ID. It highlights any canceled PMs for a specific equipment ID. This report is used to check PM compliance when a breakdown occurs. Neglecting the PM program could be the reason for the failure. This report is invaluable in justifying that PM programs are necessary to upper management.

Conclusion

This chapter does not present a complete list of reports that a manager might need to review. However, the reports presented here should be sufficient for an organization to gather enough information so that it can begin the analysis part of the benchmarking process. Without information at this level of detail, organizations are unlikely to have sufficient data to begin benchmarking.

Chapter 10
World Class Maintenance Management

What is world class manufacturing? Is it a status or a buzzword? What does the concept mean? Are other countries achieving this status? Does maintenance affect an organization's ability to reach this plateau? How can a company achieve world-class status for its maintenance organization? This chapter will answer these and other pertinent questions.

The International Marketplace

The term world class evolved from changes in the marketplace over the last several decades. In the 1960s and early 1970s, U.S. companies had both domestic and international markets. If a company chose to compete in the domestic market, it found that its competition was from within the United States as well. If it had a shortage of raw materials, a trucking strike, a rail strike, or some other interference, its competitors suffered the same problem. All other things being equal, success often came down to price, quality, and delivery. In many cases, little distinguished one vendor from another in the domestic markets. If a company chose to enter international markets, it found it generally had superior products, technology, and marketing skills compared to its competitors. U.S. companies could control the lion's share of the international market. In 1965, the United States had a 30% share of the world's market for manufactured goods, despite having less than one-tenth of the world's population.

In the late 1960s and early 1970s, a "fat cat" attitude developed; the notion that American industry could do no wrong. They could sell whatever was made, no matter the cost, no matter the quality. Although the United States was living this dream, other countries became envious. Governments, in an effort to spur their economies, encouraged economic development and these competing countries began a series of attitude changes. They analyzed what the United States was doing, and decided to do the same. As they worked 16 to 20 hours a day retooling, rebuilding, and growing their economic potential, a strange thing happened. Consider runners who have trained; they surprise both themselves and overconfident competitors who have not trained thinking they had no challenger in the field. Similarly, the market reversed. Now the Europeans and the Asian countries were making inroads, not only in their own market, but in U.S. markets as well.

U.S. companies were shocked into a state of indecisiveness. The recession of the mid-1970s reduced their ability to recover and reinvest. They began sacrificing long-term gains for short-term profits to survive. The late 1970s and early 1980s was boom time for foreign competitors. Their newfound advantage had turned into an opportunity to take market share from U.S. companies. They continued to increase quality, lower prices and deliver products just in time. They developed a competitive edge that would be difficult to overcome. The demand for foreign goods in the United States would continue to increase to the point that by 1999 the United States had a trade deficit

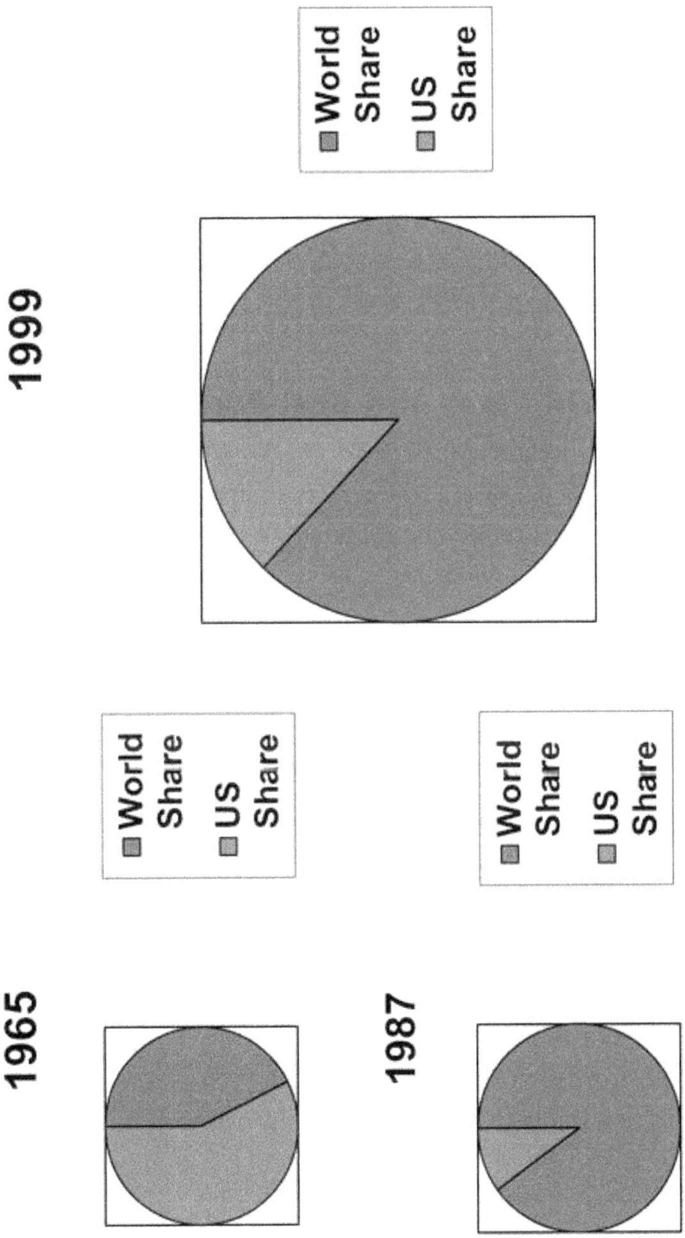

Figure 10-1, Manufacturing Market Share

of $329 billion dollars in manufactured goods. As Figure 10-1 indicates, by 1987 foreign competitors had succeeded in reducing the U.S. market share to its lowest level, 10%, of the worldwide market.

By 1999, the level had increased to 13%; still, it was still not close to the 30% share of 1965. By 2011 It had increased to 19.4% — even with that improvement, it was surpassed by China with a 19.8%. This was the first time ever in history that the United States did not have the leading percentage.

World Class and Best Practices

The true meaning of world class manufacturing now becomes clear. It is the ability to compete anywhere in the world and to be able to meet and beat any competitor anywhere in the world with product price, quality, and on-time delivery. How to achieve this level is the real challenge.

In order for a company to be World Class, it needs to implement Best Practices. However, Best Practices are not universal; they are not the same in every business market. Different markets and business conditions dictate what is Best at any given time. For example, in a market where competition dictates the price, cost control is imperative; margins are critical to control. In another market, where all the product that can be produced can be sold, capacity becomes the primary concern. Best Practices are also different for a new plant or facility than for a plant or facility that is going to be decommissioned in the next couple of years.

Regardless of the business environment, there are several universal Best Practices that can help any company improve its competitive position. These can be divided into three areas: quality, attitude towards competition, and automation technology.

Quality

A presentation at the AIPE national conference noted that 20 to 25% of a company's average manufacturing operating budget goes for finding and fixing mistakes. If the costs of repairing or replacing flawed products are included, this level may be as high as 30%. What would the amount be if it included the cost of future sales lost due to a poor reputation for quality? Estimates suggest that the cost of poor quality is as high as 10% of sales per year. Although U.S. companies have made inroads in this area, the "Made in the U.S." label has not yet regained its full appeal, especially when compared to international quality standards. The entire concept of Best Practices in quality is to prevent defects, not to correct them. Unfortunately, correcting defects is all too often the common philosophy in United States.

Attitude Toward Competition

A company's attitude toward its competitors is one of the most difficult attitudes to adjust. Moving from a domestic market to an international market, the perspective changes. Companies cannot sit back and wait for the government to impose import restrictions and try to save any industry, no matter how fundamental that industry is to the economy. If companies are to remain in business, they must focus on internal improvements, not depend on external protectionism. They must view each competitive situation as serious. The notion that "the customer will order from us because we are an American company" is based in fiction, not fact. Customers will buy where the price, quality, and service are the best.

Automation Technology

In an article published in Plant Engineering Magazine as far back as March 1978, it was stated that only 20% of the plants in the United

States were automated at the level of their overseas competitors. Consider the number of jobs that moved from the United States to overseas companies and plants in the last 30+ years. The automated plants gave companies tremendous price advantages over their under-automated competitors. (This factor is independent of the job loss related to wage differentials.) Although some inroads have been made in this area, it is commonly known that plants in the United States are still designed without full input from the operations and maintenance functions. This lack of input drives the life cycle cost of the plant or facility higher than necessary. Over the course of the plant's or facility's life cycle, the plant is forced to pay higher operational and maintenance costs. This is a large cost advantage that U.S. plants are giving their overseas competitors.

If these U.S. plants are going to lower costs and raise quality, then quality design and automation are the solution. The international companies have many industries that are totally automated from order entry to delivery, from product design to product production. A warning to companies beginning in this endeavor is that while automation is the answer, automating waste is not.

World Class manufacturing requires the elimination of complexity. It requires simplicity in design and manufacturing processes. Simplicity is the key to eliminating waste. Companies that want to achieve World Class status would be wise to examine how maintenance is involved in the areas of quality, attitudes, and automation.

Quality

If equipment is operated in an environment where nothing is done to it if it is running, then the equipment will be in poor shape. All equipment requires maintenance activities; the frequency varies with age. If equipment does not receive proper maintenance at the correct interval, it cannot maintain standards. Poorly-maintained equipment seldom produces consistently good products. If quality programs are instituted in a company, maintenance must be a part of it.

A preventive maintenance program has several components. The equipment operator may perform inspections of equipment in a world-class organization. If properly documented, these inspections can help to trigger replacements of defective components or necessary adjustments. This approach relieves some of the less technical work from the maintenance organization. It helps to promote good relations between maintenance and operations, while also helping to increase the quality of the product. Predictive and condition based maintenance systems help to insure higher quality products; the equipment condition should never deteriorate to the condition where it will be producing a defective product. Changing plant management's present mindset of "run-all-you-can" or "run-till-it-breaks" is important too, as discussed in the next section.

Attitudes

One of the attitudes that companies must address is one that favors sacrificing long term planning for short-term gains. This attitude has one meaning for the world market and the products a company plans to promote. It also has meaning for maintenance. If management's attitude favors short-term goals, then the maintenance organization will be affected in the areas of:

- Preventive maintenance
- Labor planning
- Inventory

Preventive maintenance suffers because no one will give the equipment to the maintenance department for routine service. This servicing may require some lost production, affecting short-term goals. But the company needs to view matters long range. Keeping the capital equipment in good repair and producing high quality products should be the goal of maintenance and operations. If they cooperate, long-term savings and higher product quality will result.

For example, one company, after instituting a preventive/predictive maintenance system, estimated its savings to be as high as 1.2% of the total plant output. However, the costs for the PM program start-up pushed maintenance costs up for six months before the savings could be seen. Had the company insisted in keeping costs down over the short term, it never would have achieved the greater long-term savings.

Another attitude that often requires adjustment is that of management toward the position of maintenance in the organization. Organizations that have been successful in instilling best practices in their maintenance organizations have been committed to raising maintenance management to the level of other management positions in the organization. Without this commitment, maintenance cannot improve. Foreign competitors learned this some time ago. Their organizations respect maintenance for the contributions they make to overall profitability.

Still another attitude that may need adjustment is management's attitude toward maintenance resources. These include:

- Labor
- Materials
- Tools
- Supplies and miscellaneous

Examine the maintenance budget for your organization. How much is spent annually on maintenance salaries? How much is spent on maintenance parts? You are likely to find these figures to be surprisingly low. Yet a company would probably never begin production without considering:

- The material required
- The equipment required
- The labor required
- The time required to produce the finished product

Now imagine the scenario presented in Figure 10-2. This scenario is so far removed from Best Practices that it qualifies for Worst Practices.

Most people would not want to own, invest in, or work for such a company. Any company operating in this type of mode, with this attitude, would never be profitable. It would be extremely wasteful and would create a frustrating work environment. The same feeling about a company that operates production in this manner should translate to misuse of maintenance organizations. Too many organizations operate their maintenance organizations in a way consistent with Figure 10-2. The reason is due to shortsighted management, who place no importance on maintenance. Without a change of attitude, these types of maintenance organizations have little or no chance of improving.

FIGURE 10-2 "WORST PRACTICE" OPERATIONS
▶ Each operator waiting for someone to tell them which equipment to operate? ♦ Or the operator telling production management which product they are going to make today?
▶ Each operator going to the warehouse to get the material needed to make the product as the production process required it?
▶ Each operator ordering their own material to make their assigned product and then waiting for it to be delivered?
▶ Each operator waiting in line to use a piece of equipment to make their assigned product because other operators needed the same equipment at the same time?
▶ Operators standing around, watching an operator work, because no one knew that it took only one operator to run the machine, and multiple operators were scheduled for the equipment?
▶ Ridiculous?

Figure 10-2, "Worst Practice" Operations

Maintenance planning is the most essential part of any company's effort to permanently improve the efficiency an effectiveness of the maintenance organization.

In short, management must understand maintenance management if it is to improve. If they do, they can reap the following benefits:

- Quality improvements
- Improvements in utilization and uptime
- Reduction of maintenance labor and material costs

They will also have an organization that is prepared to do business in the future world market, no matter how severe the competition.

Automation Technology

To the maintenance organization, automation technology means being able to take advantage of all of the technological advances in maintenance tools and equipment in order to keep maintenance costs at an optimum level. For example, consider vibration analysis. When it began, it was strictly an engineering function. As the technology matured, the tools became easier to use. Soon, maintenance technicians were regularly involved in making and trending the readings. Complementary software was then developed that made the charting, trending, and interpretation of the date even easier. Currently, some companies have "go/ no-go" indicators mounted on the equipment and operators can quickly spot problems and relay the information to the maintenance organization for resolution. It is the exception to find a manufacturing company today that is not involved in some form of vibration analysis. This predictive technique alone has provided companies innumerable savings.

Other areas of automation technology include additional predictive maintenance techniques such as infrared scanning and spectrographic wear particle analysis. Although these techniques are only recently coming into wide use, they have the same potential as vibration analysis to improve maintenance performance and utilization. Another more advanced tool is condition-based maintenance, in which real time inputs are fed to maintenance so that the parameters can be analyzed and appropriate action taken. It will be several years before this technology is widely accepted. The computerization of maintenance information is another example of automation technology. Computerized Maintenance Management Systems (CMMS) or Enterprise Asset Management (EAM) Systems automate the paper flow used by the entire maintenance organization. These systems consist of several sub-systems, including:

- Equipment data
- Preventive Maintenance
- Work orders
- Inventory
- Purchasing
- Personnel
- Reporting

Using computers to automate the maintenance information gives management access to historical data, concise summary reports, and, in most cases, graphic display of the data. However, some companies bury themselves with the amount of data they produce with their CMMS or EAM systems. Information should support management, not burden it.

CMMS or EAM Systems are often the final key that allows maintenance to take full advantage of automation technology. Vendors, consultants, and software experts pull information from predictive maintenance systems, computer-based training systems, and other related packages into the CMMS/ EAM System. One example uses the predictive maintenance information to trigger repair work orders in the CMMS/ EAM System; the system plans and schedules the work orders according to their priority. All of these possibilities are important because they again reflect the impact that maintenance has on corporate profitability.

ROFA, Maintenance Management, and Asset Management

The investment a company makes in its assets is often measured against the profits the company generates. This measure is called return on fixed assets (ROFA). This indicator can be used in strategic planning when a company picks what facility to occupy or the plant in which to produce a product.

Asset management focuses on achieving the lowest total life-cycle cost to produce a product or provide a service. The goal is to have a higher ROFA than competitors, so as to be the low-cost producer of a product or service. A company in this position attracts customers and ensures greater market share. Also, a higher ROFA will attract investors to a company, ensuring a sound financial base on which to build further business.

It is the responsibility of all departments or functions within a company to measure and control their costs, because ultimately they will impact the ROFA calculation. Only when all departments or functions within a company work together can the maximum ROFA be achieved. It is beyond the scope of this text to deal with all these areas in detail. The focus here is the maintenance function.

Maintenance and Asset Management

In what ways does maintenance management impact the ROFA calculation? Two indicators may be used to show the impact:

Maintenance costs as a percentage of total process, production, or manufacturing costs.

This indicator is an accurate measure for the costs of manufacturing; it should be used as a total calculation, not a per-production-unit calculation. Maintenance will be a percentage of the cost to produce, but is generally fixed. This stability makes the indicator more accurate for the financial measure of maintenance because it makes trending maintenance costs easier. If the maintenance cost percentage fluctuates, then the efficiency and effectiveness of maintenance should be examined to find the cause of the change.

Maintenance cost per square foot maintained.

This indicator compares maintenance costs to the total amount of floor space in a facility. It is an accurate measure for facilities because the cost is also usually stable. This indicator can easily be used to trend any increases over time. If the percentage of maintenance costs fluctuates, then the efficiency and effectiveness of maintenance should be examined to find the cause of the change.

These two indicators show that traditional maintenance labor and material costs will have an impact on the ROFA. However, insuring the equipment or assets are available can also have an impact. In this regard, there are two main areas to examine: 1) maintenance costs and 2) equipment or facility availability.

Maintenance Costs—Labor

Maintenance productivity in most companies with reactive maintenance policies averages between 25% and 35%. These percentages indicate that fewer than three hours per eight-hour shift are spent on hands-on activities. Most of this lost maintenance productivity can be categorized into the following kinds of delays:

- Waiting for parts
- Waiting for information, drawings, or instructions
- Waiting for equipment to be shut down
- Waiting for rental equipment
- Waiting for other crafts to finish their part of the job
- Running from emergency to emergency

Although 100% maintenance productivity is an unrealistic goal for any maintenance organization, 60% is achievable.

The productivity of maintenance technicians can be improved by concentrating on basic management techniques, such as:

- Planning jobs in advance
- Scheduling jobs and coordinating schedules with operations or facilities
- Arranging for parts to be ready
- Coordinating tools and rental equipment
- Reducing emergency work to below 50% (measured by work orders)

With computer assistance, planning time per job is reduced, resulting in more planned and coordinated jobs. More time is then available for preventive maintenance activities which, in turn, help to reduce the amount of emergency and breakdown activities. The results include fewer schedule changes and increased productivity (by reducing travel and waiting times). Organizations that are successful in achieving good maintenance labor controls experience significant increases in labor productivity.

Maintenance Costs—Materials

Material costs are related to the frequency and size of the repairs made to the company's assets. The sheer number of parts, in addition to stores policies, purchasing policies, and overall inventory management practices, contribute to overall costs of maintenance materials. In some companies, little attention is paid to maintenance materials, and inventories may be higher than necessary by 20% or 30%. This level increases inventory holding costs and makes materials unnecessarily expensive. Sometimes the inability of stores to service the maintenance department's needs results in "pirate" or "illegal" storage depots of just-in-case spares. This practice also drives up the cost of maintenance materials.

Good inventory controls enable companies to lower the value of the inventory and still maintain a service level of at least 95%. Such levels enable maintenance departments to be responsive to the operations or facilities groups, while increasing their own personal productivity. Organizations that are successful in managing their maintenance inventories typically average 19% lower material costs and an overall 18% reduction in total inventory compared to companies that have not focused on this area.

Equipment or Facility Availability

Consideration of equipment or facility availability reveals the connection between asset management and maintenance management. Downtime cost for equipment may vary from several hundreds of dollars per hour to literally hundreds of thousands of dollars per hour. These costs are due to lost production from assets and lost or reduced efficiency (or occupancy) of a facility. In some companies, levels of downtime run beyond 30%. Such levels result in lost sales opportunities and unnecessary expenditures for capital equipment. In general, the organization is in a weak competitive position.

By committing the organization to good maintenance policies and practices, and by using its computerized maintenance management system as a tracking tool, management can reduce equipment downtime. The result is more throughput, and more throughput enables the company to get more products or services from its assets, resulting in lower production costs and a higher ROFA.

Maintenance and ROFA

If asset management is a focus for an organization, the maintenance function can contribute to overall plant profitability. The cooperation and focus of all departments and functions within an organization are needed in order to be successful, but the maintenance department can have a dramatic positive impact on ROFA.

Because maintenance is typically viewed as an expense, any maintenance savings can be viewed as directly contributing to profits. By achieving maximum availability from equipment, plant or facility managers insure that a company does not need to invest in excess assets to produce its products or provide its services. This result is a good indication that a company is truly managing its assets.

Technology will continue to advance and will be used to further optimize maintenance resources. World-class organizations will take advantage of these advances. They will have to because their competitors, both internationally and domestically, will. Procrastinators will lose because Best Practice companies are proactively seeking this competitive advantage.

Conclusion

This chapter has looked at the competitive benefits of Best Practices maintenance. The impact maintenance has on a company's ability to compete was also examined. Some of the benefits of implementing Best Practices in maintenance were also considered. But ultimately, this chapter has highlighted attitudes. One of the major reasons companies fail to achieve Best Practices in maintenance is their overall attitude about maintenance. Companies that understand how the maintenance function impacts their competitive position will be in a better position to achieve maintenance Best Practices.

Chapter 11
Integration of Maintenance Management

In today's competitive markets, companies must investigate every potential business opportunity that might turn into a competitive advantage. Companies that make the most of the tools they use to do business will be dominant in their respective marketplaces. One set of tools that bears close examination is information systems, the computerized tools that companies use to help monitor and control their businesses.

A discussion of every kind of information system available to a business is beyond the scope of this text. However, space does allow descriptions of several common information systems and their effective uses.

Typical Systems

The most common systems used in business today include the following:

MRPII and MES Systems

MRPII and MES systems help companies manage the financial, manufacturing, and distribution aspects of their business. Manufacturing resources planning (MRPII) systems are evolving into manufacturing execution systems (MES). In the future, MES systems likely will evolve into enterprise resource planning (ERP) systems, the integration of all aspects of a business into a single, comprehensive system.

CMMS and EAM

CMMS (Computerized Maintenance Management Systems) and EAM (Enterprise Asset Management Systems) help companies manage the condition of the capital equipment they use to produce products or provide services. These systems can track information about the equipment and also schedule routine and preventive maintenance. Using maintenance information, companies can make objective repair-or-replace decisions and ultimately track total life-cycle costs of capital assets.

Shop-Floor Control Systems

Shop-floor control systems help companies plan and track the progress of products through a manufacturing process. These systems allow reliable forecasting of production times and projected delivery times, both of which are required for effective production in a just-in-time (JIT) environment.

CAD/CAM Systems

CAD/CAM systems (computer-aided design and computer-aided manufacturing) support the design and manufacture of a company's products. Because they allow manufacturers to reduce the time required to introduce a new product into the marketplace, these tools are extremely valuable to industries that manufacture products with short life cycles. That is, they allow a company to speed up the time to market for their products.

ENERGY MANAGEMENT SYSTEMS

Energy management systems allow companies to manage and control building or process systems that use energy. These systems may control heating, ventilating, and air-conditioning (HVAC) systems; lighting systems; process heating and cooling; etc. Optimizing the control of energy-consuming systems lowers energy costs. Companies often combine security and fire-safety systems into energy management systems. When this occurs, these systems are sometimes called building management or building automation systems.

Effectively Using Information Systems

One measure of the effectiveness of an information system is how much data can be accessed and used by those needing the data. The more effectively these systems are used, the greater are the benefits, returns on investments, and ultimately a company's competitiveness. In practice, any one of these systems may function in one of four ways:

- As a stand-alone system
- To batch transfer and receive data from another system or other systems
- As a system interfaced with other systems
- As a system integrated with other systems

This list represents a progression from adequate to optimum use of data in a computerized environment.

Stand-Alone System

Stand-alone use of an information system usually indicates that a department controls its use. Data collected stay under the control of the department that "owns" the system and is not easily shared with other departments. The department in control of the system uses the system to manage or control its business. At the same time, that department may be oblivious to ways the information could help meet the needs or operational concerns of other departments in the company.

Batch Data-Transfer

Batch data-transfer means that one department has some level of recognition of the information requirements of another. Data transfer may be as basic as running a report out of one system, then carrying it down the hall for re-keying into another system. A higher level of sophistication might involve downloading a text file or a standard database format to a disk or tape and then uploading it into another system.

Interfacing Systems

Interfacing one information system to another is the next step above data transfer in increasing the effectiveness of information systems in order to achieve a competitive advantage. This phase often takes place when the daily transfer of data between systems becomes tedious and electronic transfer promises to increase operational efficiency.

Interfaced systems are periodically forced to "talk to each other" or, in some cases, programmed to share data automatically. Interfaced computers, operating as described, pass the required data in batches. This means that at a fixed time, data from one system updates all related records in another system. Although this mode of operation has distinct advantages over the manual transfer of data, it also means that between updates a user or manager might be looking at old or incomplete data. In today's competitive environment, delays or decisions made on the basis of old data can be costly.

INTEGRATING SYSTEMS

Integrated information systems pass data among themselves in real time as it happens. The most advanced information systems have integrated databases so that the various dedicated systems (CMMS, energy management, etc.) are actually parts of one large system. In these systems, all parts of the system that need information share it. Integration ensures ease of data manipulation and the timeliness of the data. It also enhances the optimum use of the data in a competitive environment.

TWO TRAPS

Information systems present several traps for plant and facilities professionals. Two of the most prevalent traps entail the following beliefs.

The first trap is the belief that, as a package, information systems will solve any problem. Some companies have become so enamored with technology that they attempt to buy an off-the-shelf solution to all of their problems. However, before companies attempt to automate their business processes, they should understand those processes and what the processes should be. In the absence of an understanding of the discipline required to effectively execute a company's basic business functions, the installation of an information system to automate those functions will likely fail. Systems only provide and manipulate information. Companies must enforce the discipline required to make the systems effective.

The second trap is believing that the MIS department (management information systems) is always the best source of information about what hardware and software a company needs. When a company starts to evolve from stand-alone systems toward integration, this attitude can result in failure to achieve the desired results. Because the MIS department is typically responsible for managing a company's computer systems, it tends to take ownership of the evolving, integrated system. Instead of allowing other departments to select systems that support their philosophy of doing business and, then integrating their selections, the MIS department too often dictates, based on its own software and hardware preferences, what system must be used.

All departments involved in the use of information systems must play a role in their selection and integration. When this does not happen, successful integration seldom occurs. Then, evolution toward integration usually becomes regression back to a collection of stand-alone information systems.

Integrating Maintenance Into The Business Plan

After reviewing the types of information systems available and understanding the competitive benefits of integrated systems, a company must ask itself this question: Are we taking advantage of the best computer tools available to us today to make our business more competitive in the future? The answer to the question begins with attitude and focus. Poor attitudes about maintenance have been discussed previously in this text. However, if maintenance systems are not given the same priority in the company as other business systems, the data will never be accurate enough.

CIM (Computer Integrated Manufacturing), MRP (Material Requirements Planning), MRP II (Manufacturing Resources Planning), JIT (Just in Time), FMS (Flexible Manufacturing Systems), CAD/CAM (Computer-Aided Design/Computer-Aided Manufacturing), MES (Manufacturing Execution System), and ERP (Enterprise Resource Planning) are all acronyms that are commonplace in manufacturing today. Yet in all the literature on these topics, almost no mention is made of the maintenance organization or the interaction that maintenance management must have with other parts of the organization to make these systems successful.

The problem is one mentioned earlier: Attitude. Engineers and managers who conceive and implement these systems forget a set of basic concepts. To manufacture the product you need equipment. To operate in a world class manufacturing organization, you need automated equipment. To operate automated equipment, you need automated control systems. To insure that the automated control systems and the automated systems function, you need a Best Practice or world-class maintenance organization. If maintenance is not included in a business plan for automated manufacturing, the plan will not be successful in the long term. Although this last statement may seem to be rather bold, it is true. Yet some companies will point to a new plant or facility, and suggest that maintenance is not necessary; the plant or facility operates relatively free of maintenance.

How can this last statement be true? It is true only because all of the equipment is new. It has not aged to the point where maintenance is required. Depending on the process and loads put on the equipment, the condition deteriorates in three to five years. Breakdown rates climb, while utilization rates drop. Some companies decrease the yield of the equipment to compensate for the amount of maintenance-related equipment downtime. No matter how companies try to cover over the fact, maintenance is an important and necessary part of any world class manufacturing organization.

Maintenance and MRP

How does maintenance fit into the total picture? Consider this example. MRP systems use the master production schedule and the bill of materials to determine what equipment will be required, what parts are necessary and what labor is needed to produce a product. The master production schedule has all of the sales forecasts, allowing the output for the company to be known. The bill of materials helps to determine what the total materials demands will be from stores. The production scheduler knows how long each item takes to produce on each piece of equipment, allowing the labor resources to be determined. This process seems simple enough and can be very accurate. But the inaccuracy in the system develops when the assumption is made that the equipment will be available when necessary to produce the product.

If the schedule requires the equipment to run for 16 hours a day for 5 days during the week, the equipment must run at standard operating speeds, producing a quality product for 80 hours. Suppose the equipment breaks down, stopping production for 4 hours on three separate shifts, or 12 hours total. In addition, during another 8 hours, the hydraulic system (a part of the equipment system) would not develop enough pressure to allow normal operating speeds and the equipment had to run at 50% speed, adding another 4 hours (8 x 50%)

of lost production. Thus, there is a total of 16 hours of lost production.

How would this lost production be made up? There would be two options: run two extra shifts on the sixth day, or push the added demand into the next week's schedule. Pushing the orders back into the next week is not an acceptable solution to the JIT environment. It would delay the order, possibly not allowing it to arrive at the customer's site in time for their needs. Such a delay can result in the loss of a customer in today's competitive marketplace. The only acceptable solution is to operate the two extra shifts. This option enables the production to be made up with no appreciable delay.

But what costs are involved? Here is a partial list:

- Overtime production labor
- Overtime maintenance labor
- Extra utilities to run equipment
- Extra time to re-do the production schedule
- Extra time to inform customers order may be delayed

The costs for these extra shifts could have been avoided with an effective maintenance program. Most experts agree that the only way most new production methods will work is with a rigid preventive maintenance program. In this last example, suppose a preventive maintenance program had been in place. If the PM program included newer technologies, such as vibration analysis, the causes of the breakdown would probably have been detected before the breakdown occurred. The repairs could then have been scheduled for the off shift during the week, eliminating the delay for the production unit. The clogged filter, which reduced flow on the hydraulic system, with the subsequent loss of pressure, would have been replaced during an off shift as part of the routine PM service for the hydraulic system. This step alone would have eliminated the lost production time due to reduced speed operation.

Although this example may seem simplistic, it has actually occurred at some plants. Every company has similar stories to tell. All include unnecessary breakdowns that could have been avoided with a management structure that gives importance to maintenance management. How does the maintenance function and its information system fit in the automated factory and the JIT environment? Through the four options that were discussed previously: basic, batch, interfaced, and integrated.

The Basic Option

A basic stand-alone solution is manually input from a maintenance program into the production scheduling process. The maintenance manager compares schedules with the production manager to look for any conflict between the time that maintenance requires the equipment for service and the time that operations needs to make product. Any conflicts can then be resolved on a case-by-case level, allowing the decision to be made to risk a problem or to correct it. The plant manager should resolve any decision that the maintenance and operations managers cannot.

The Batch Option

A batch-loaded system is typically used between a maintenance system and an inventory system, especially when the two business functions (maintenance and inventory) can not agree on using the same system. Therefore, each uses its own system. Then at a specified time interval, the stores catalog databases are synchronized, allowing for the data to be accurate enough, except for the most demanding of environments.

The Interfaced Option

An interfaced system is used when the corporate systems (production planning, financials, purchasing, etc.) and the maintenance management function are computerized. The two systems can operate independently of each other except when information is batch loaded from one to another. The most common example is loading the maintenance demands for time to the production scheduling part of the production planning system. The production planning system treats the maintenance demands for the equipment just as if they were product demands, allowing for a smooth scheduling process. If there are no conflicts, the schedule can be produced and finalized. If there are conflicts, they can be handled by shifting resources, off-loading production to other equipment, or by postponing the maintenance request (if not urgent) to the next week. The two systems can work together, avoiding the "islands of automation" problem.

The Integrated Option

The most advanced system, which is just now beginning to be used, is an integrated system. It is different from the interfaced system because it is real time, not batch loaded. This type of system will become even more important when the PLCs on the shop floor are feeding back information into the production scheduling and maintenance systems in a real-time environment. This condition-based environment is one that maintenance organizations will have to move to in the future if they are to contribute to corporate profitability in a world-class organization.

In this organization, all production information is fed back into the production scheduling system for compliance checks against the master production schedule. Part of the same information (run times, production rates, etc.) is also fed into the maintenance system. This information, coupled with the data from vibration, temperature, or sonic sensors, is used to schedule maintenance Just-In-Time to prevent equipment outages or quality problems. This synergistic relationship will be necessary for the company to achieve optimum cost-product-service relationships. Only by achieving this type of cooperation will any company be able to remain competitive in the world market. If companies delay adding maintenance to their total plan for organization improvement, the world market will soon pass them by.

Enterprise Asset Management

Astute observers of the market for computerized maintenance management systems will have noticed that software vendors are now calling their products enterprise asset management (EAM) systems instead of CMMS. To understand the reason for this change, one must be aware that many companies use enterprise resource planning (ERP) systems to manage all resources required to produce a product or provide a service — from material receiving through production through shipment to customers. These systems are connected with businesses from order entry to order fulfillment.

By contrast, CMMS is used by maintenance departments to manage the maintenance function. Typically, a CMMS is independent of the main business system, requiring manual schedule integration to avoid conflicts.

When conflicts arise between ERP systems and CMMS, they are often because of the failure to give sufficient emphasis to the maintenance function. In general, a company cannot successfully plan resources (assets) at an enterprise level without managing assets at that level. Conflicts develop when, due to poor equipment availability, excess assets are purchased to ensure enough capacity to meet market demands. This strategy is uncontrolled, often resulting in excessive maintenance, repair, and operating costs as well as lower asset utilization. Excess (under-utilized) assets lower the return on all assets, signaling a poor investment.

The solution is a move beyond maintenance management to EAM. EAM systems seek to manage a company's assets to optimize their use, thereby maximizing the return on investment in the assets. EAM includes using in-process information in a "health analysis" designed to deliver just-in-time maintenance with production impact included in the optimization equation. In other words, EAM takes a process- or asset-centric view of the enterprise, as opposed to a product-centric view.

In brief, then, ERP systems entail planning based on capacity. EAM enables or delivers that capacity. Thus, EAM is more than maintenance management, and EAM software aims to be more than maintenance management software.

Conclusion

Maintenance has been called the last frontier for management to conquer. Just as there have been pioneers in the past who went out and explored the frontier ahead of the crowd, so too have there been pioneers in the frontier of maintenance management. They have been able to take advantage of the cost benefits available. You are able to read of their exploits in magazine articles or listen to them at conference presentations. Just as pioneers of old would return to tell their tales, spurring others to move into the frontier, so we can hope that the pioneers of maintenance will inspire others to move into the last frontier for manufacturing and facilities. Problems will develop, but overcoming them brings a sense of pride and accomplishment.

The final question this book asks is "Are you and your organization up to the challenge?" Innovators are making things happen. Procrastinators will not conquer the final frontier. The choice of which group you belong to is up to you and your organization. The choice will have lasting effects on your survivability. The information in this book should help you make the correct decision.

Chapter 12
Benchmarking Best Practices in Maintenance Management

When benchmarking, it is best to determine early in the process what you are trying to accomplish. Are you really trying to improve or are you just trying to find some number to reach? If you are truly trying to benchmark, then self-assessment is an important prerequisite.

Finding Benchmarking Partners

One of the most difficult challenges when benchmarking is finding legitimate partners. Although a variety of articles and Internet sites might suggest potential partners, it is important to find a legitimate benchmarking partner, one who has documented best practices in a particular process of maintenance management. The benchmarking partner should have had the desired process under control for a period of time and should also show sustainable results over the same time.

Once you identify the benchmarking partner, you must determine what you have to share with the partner. For you to have an interchange of ideas, you must have a best practice that you can share.

This practice does not necessarily need to be superior, but it should be something of interest to the partner. Otherwise, the benchmarking exercise is a one-way learning experience and partners can quickly sour on the experience.

During the benchmarking process, key data about best practices will be identified; this data should only be shared among those participating in the study. Sharing the data with someone outside of the benchmarking study is both unwise and unethical, and can turn many companies away from benchmarking.

Variables in Benchmarking

Achieving accurate comparisons requires both data and a clear understanding of the processes and parameters that are being measured. Only then can there be an apples-to-apples comparison. This section examines some of the variables that can be challenging to the benchmarking process.

Equipment Availability

One of the first challenges is identifying the true meaning of equipment availability. Is equipment availability measured only when the equipment is required to run but does not run? How does idle time factor into the measurement, when the equipment is not running, but is not scheduled to run. How is equipment availability different in a market that is sold out, where capacity is deficient, compared to a marketplace which has excess capacity, providing the company the ability to work around the equipment being down? Unless these issues are clearly defined and understood, any benchmarks that would be derived are meaningless.

Planning and Scheduling

Another challenge is defining planning and scheduling efficiency. The definition of planned is itself extremely important. How far in advance must a job be identified and scheduled for it to be considered as planned: 30 days in advance, a week, or some other length of time? In some instances, companies determine a job as planned even if it has only four hours notice. This short-term planning window will not improve the cost-effectiveness of performing maintenance. The planning window should be at least seven days to be considered planned. If a company that uses a short-term planning window benchmarks with a company that uses a long-term planning window, the comparison is meaningless.

Comparing planning efficiency — what is planned — can be challenging. Are contractors, or equipment and tools, part of the plan? Does the plan include labor and materials? Is the plan penalized if unplanned graphs or unplanned materials used to perform the job? These issues must be discussed during a face-to-face meeting with a benchmarking partner. They are factors that do not show up when examining only benchmarking numbers.

Another consideration is whether or not preventive maintenance activities are counted as planned work. Companies that include preventive maintenance as part of planned work have higher planned work percentages. When preventive maintenance activities are not counted, planned work percentages are lower. Understanding what is actually included as part of the planned work calculation is important.

Inventory and Purchasing Benchmarks

Another benchmarking challenge is to understand the inventory and purchasing numbers. For example, what defines a stock out? Is an item considered out of stock if it is not available when the job is planned? Or is it only considered a stock out if the part is not available for issue when the job is ready to be executed? The difference between these two definitions makes an enormous difference between the two stores service levels calculations.

Another typical stores benchmark is the calculation of the total amount of dollars tied up in the stores investment. The lower the dollar value of the spare parts a company carries, the lower the stores investment number. Conversely, a higher value of spare parts that a company carries in stock leads to a higher stores investment number. Some parameters automatically penalize companies. For example, companies that use a lot of foreign equipment must stock a larger number of spare parts. If they do not overstock these parts and an equipment breakdown occurs, the lead time to get the spare parts could be extremely long. In fact, some equipment suppliers only make spare parts during one time of the year. Therefore, if a company uses foreign equipment, it is required to keep more spare parts. Even though this higher inventory is legitimate, it will inflate the stores investment.

Another factor that has a dramatic impact on the stores investment benchmark is the relationship that companies are able to develop with suppliers. In large metropolitan areas, companies may actually have consignment relationships with suppliers. In these cases, the supplier will actually stock spare parts for the company, which does not own the spare parts until it actually uses them. This relationship is much more difficult to develop in remote rural areas. Suppliers are not close to the plants and the downtime that would be required to procure the spare part and then have it delivered to the plant is cost prohibitive. Unless these factors are clearly understood, benchmarking numbers is meaningless.

A third factor is the amount of spares that a company may capitalize. In some companies, major spares are capitalized as part of the overall equipment acquisition costs. These spares will no longer show up with an attributed value on the inventory. If a company does not capitalize these types of spares, their value of inventory on hand will be much greater than a company that did capitalize the spares. This is a "hidden factor" that must be investigated if there is a large difference in the value of spares between benchmarking partners.

Maintenance Labor Resources

Another challenge is understanding what is calculated as part of the maintenance labor resources. Are only the actual craft technicians counted as maintenance labor? Or are all those who perform some maintenance activities translated into full-time equivalents and then included as maintenance labor? Because they are paid from a separate budget, are contractors calculated as part of the maintenance labor force? There can be a large difference in how different companies calculate manual labor.

Similarly, comparing the cost of operator-based maintenance can be challenging. To calculate this amount of maintenance properly, companies should convert the operator-based activity hours into full time equivalents and add their hours to the maintenance labor costs. If a company does not do this, but its benchmarking partners do, they would face a large difference in the amount of maintenance labor costs needed to staff the plant.

Maintenance Cost Comparisons

Total maintenance costs are a measure commonly used for benchmarking. They may be compared to a production parameter or to some type of square-foot measurement. Without understanding the way in which the maintenance organization operates, this measure will be used incorrectly. Reactive maintenance activities usually cost between two to four times more than proactive maintenance activities. Therefore, a proactive maintenance organization automatically gains when comparing costs to benefits. However, if a proactive organization is compared to a reactive organization on the basis of costs only, senior management may determine that maintenance costs are too high and immediately begin ordering cuts. To improve the maintenance cost benchmark, increased spending is required to move an organization from a reactive to a proactive position. After proactive status is achieved, the maintenance costs should normalize when compared to the benchmark. However, if senior management makes cuts prematurely, the needed maintenance resources will not be available and the desired benefits will not be realized.

Another way to look at maintenance costs is to compare labor costs to material costs. Standard comparisons should show maintenance and materials each with 50% of the costs. In certain industries, labor could be 60% and materials 40%, or the reverse. However, if there is a deviation to a 70 to 30% ratio, then perhaps too many parts are being changed or a lot of labor is nonproductive. Without this comparison, a simple look at maintenance expenses would be incomplete. An initial analysis of your organization and a detailed analysis of the benchmarking partner's organization are both needed before any true benchmarking can be valuable.

Another maintenance process that can have a tremendous impact on the cost ratio is the effectiveness of the planning and scheduling activities. The more effective that planning and scheduling activities are in the maintenance organization, the lower the overall maintenance costs will be. Conversely, the more reactive that the maintenance organization is, the higher its overall maintenance costs.

Unplanned work will cost two to four times as much as planned work. The organization that is efficient at planning and scheduling maintenance activities will receive a cost-benefit. Conversely, a maintenance organization that is reactive automatically incurs an additional cost penalty. The efficiency of the planning and scheduling program must be examined during any maintenance benchmarking.

Preventive Maintenance

Another area in maintenance that is often benchmarked is the preventive maintenance program. However, this benchmarking is not as easy as it may seem. What is considered a preventive maintenance activity? In some companies, preventive maintenance programs include repairs that are needed while doing a preventive maintenance inspection task. In other companies, preventive maintenance is only an inspection service. Once the inspection is made, work orders are written, and then the repairs are planned, scheduled, and executed. Some preventive maintenance programs that perform minor repairs during the inspection will experience lower costs than an organization that merely does an inspection and then comes back and writes the work order to have the service performed.

However, if an organization performs excessive repairs during preventive maintenance inspections, then its overall preventive maintenance costs will be higher than an organization that limits itself to a fixed amount of maintenance repairs. It is not the purpose of this discussion to say which method is right or wrong. It is only to highlight that unless the operating policy for preventive maintenance activities is clearly understood, getting an accurate benchmark comparison will be difficult.

Hidden Factors in Benchmarking

This section considers hidden factors that enable a company to achieve superior numbers in benchmarking. However, these factors are not discernable without close scrutiny. Each should be closely examined during any benchmarking project.

Organizational Impact

When examining maintenance organizations, it is necessary to understand several factors. The first is the deployment model. Should the organization be centralized, decentralized, or a combination of the two? If an organization is in the wrong deployment model, its costs will be higher either in wasted labor resources or excessive equipment downtime. Furthermore, if the organizational deployment model is not clearly tied to the maintenance business plan, it will be difficult to match the results of one benchmarked organization to another.

A second factor is the reporting structure. As was discussed in Chapter 3, there are multiple ways for maintenance to report within an organization. The organizational reporting structure has a dramatic impact on overall maintenance benchmarks. Therefore, the organizational reporting structure must be clearly defined and understood among the benchmarking partners before any meaningful comparison can be achieved.

No organization that employs an operating team strategy or a business unit strategy of distributed maintenance is among the low-cost producers in any market. When operating team or business unit approaches are utilized, the distributed maintenance model results in redundant resources and excessive costs. If this model is in use by any company involved in a benchmarking project, a close examination of the model's impact on any statistics is needed.

Management and Employee Attitudes

Employee attitudes are extremely important when considering benchmarking data. The overall attitude of the salary and hourly employees is critical to achieving Best Practices. One survey showed that employee's attitudes impacted the benchmarking results more than any other factor. Suppose a plant has an attitude that maintenance is a necessary evil, and not a contributor to overall profitability. The benchmarking data will differ significantly from another plant where maintenance is considered to be a contributor to profitability and a valued partner in operating the plant.

There are also significant differences in benchmarking results at plants where there is an adversarial union-management relationship. When comparing this type of the plant to one with a cooperative attitude, the results from benchmarking maintenance will differ. Yet in many benchmarking studies, this factor is never clearly presented. To gain any insight into the enabling factors behind a benchmarking statistic, this area must be examined.

While considering employee attitudes, it is also necessary to examine management attitudes. Senior management's understanding of maintenance management varies dramatically from plant to plant. In plants where senior management understands maintenance and manages it correctly, maintenance efficiency and effectiveness allows the plant to be competitive. However, in plants where senior management does not understand maintenance management, the resources are squandered and the overall maintenance cost is much higher when compared to other plants. In many cases, senior management is seen throwing resources at problems, rather than understanding and correcting the root cause of the problem. This has led to excessive expenses in the maintenance function with little or no long term improvement. It is important when benchmarking different plants to understand the senior management perspective on maintenance. This perspective can be a factor that will enable or disable the maintenance function.

Studies have shown that organizations with the highest maintenance costs typically keep their technicians busy repairing failures. At these plants, the technicians have no opportunity to take time to examine the root causes of failures; they are unable to take the time to make the repairs permanent or devise a preventive and predictive solution when the problems can't be prevented. From a financial perspective, these plants are typically the worst performers from a financial perspective.

Plants tend to divide into two categories. The first is a repair-focused organization, one that assumes equipment will fail and maintenance's mission is to respond quickly to equipment in distress. This type of plant is typically known as a firefighting, or reactive maintenance organization. It has no opportunity to examine failure causes, but rather focuses on just "fixing it." When maintenance is not busy fighting fires, it focuses on low priority work to appear busy. This method is used to protect an already overworked staff.

The second category is a reliability-focused organization. Here, equipment breakdowns are not expected to happen. When they do, they are viewed as an exception and typically result from some flaw of maintenance policy or management focus. The vision of re-occurring failures and the related costs is deemed unacceptable. Reliability-focused organizations are always a low-cost producer in their respective marketplaces. Before beginning a benchmarking exercise, it is necessary to understand which approach an organization has as well as its philosophy toward equipment failure.

Additional Best Practice Enablers

The management at Best Practice companies purposefully manages reliability for results. They make permanent repairs when needed, but relentlessly assess equipment condition. Thus, they are continuously analyzing plant data. Their decisions are based not on guesswork, but on actual plant data, allowing them to make optimal financial decisions based on real-time data, not assumptions.

Another management enabler is the recognition that plant reliability is not a repair effort. Maintenance is not status quo. Instead, maintenance focuses on eliminating the root cause of the failure. Any less effort is not acceptable in a Best Practice organization. Maintenance is tied to improvement and optimization.

Another enabler for Best Practice companies is the operating information and its value. These companies recognize that data is a company asset in which they have invested; this data must be utilized, both at the current time, and in the future. These organizations do not accept excuses for not recording data, nor do they accept excuses for not using the data when planning future activities. Informed decision making eliminates the "I think" and "I feel" syndromes. All activities are financially justified based on data.

Accountability is also a management enabler. Best Practice organizations view maintenance as a core business process. They detail a three-to-five-year business plan for maintenance. They set reliability performance targets, organizational objectives, maintenance budgets, reliability improvement goals, and spending to improve equipment. Once these goals and objectives are set, the maintenance manager is held accountable for achieving results. If accountability is absent, then cost-effective, organized problem-solving and results compared to budgets are never observed. In best practice companies, the maintenance manager is a key position and must be staffed by an appropriate person.

A practice that is being implemented in many best practice companies is operational ownership of the equipment or process. Instilling equipment ownership in operating personnel helps the maintenance department optimize equipment effectiveness. This practice — increasing operating awareness to their role as equipment owners — improves equipment and plant capacity; it also has a significant impact on profitability. When benchmarking any company, examine this area because it will have an impact on any benchmarks.

As far back as 1989, a summary of the Best of the Best Maintenance Organization Award (Sky Magazine, September 1989) highlights areas that are still as relevant today when benchmarking.

1. The Best do the basics very well; but even among the Best, there is still room for improvement. Studies have shown that up to half of all plant equipment failures can be traced to the neglect of basics. If the Best focus on the basics, then shouldn't all organizations? It will be apparent when benchmarking with a Best Practice company that the basics are given great attention. This is an area that should be considered as a hidden enabler when benchmarking a Best Practice company.

2. No significant advances in maintenance management technology were found in the Best companies. This observation is important and links directly to the first observation. There is no magic black box that will make an organization the best. Focusing on the basics and giving attention to detail will help differentiate a company and make it the best. No amount of technology can do that. Only with focused management, as well as a highly motivated and empowered work force, can this occur.

3. Individual leadership was the most common factor among the best. Motivated leadership with a focus on maximizing equipment effectiveness is crucial in producing results that will qualify a plant to be the best. In organizations that are recognized as the best, a common element is always a motivational and business-oriented maintenance manager. Without this individual, it is difficult, if not impossible, to achieve Best Practice results.

Best Practice Benchmarks: Asset Value-Based Benchmarks

One benchmark that is rapidly gaining acceptance divides the total maintenance cost by the estimated replacement value (ERV) of a plant or a facility. The maintenance costs are easily derived from either the maintenance budget or accounting. The estimated replacement costs of the plant or facility are more complex. However, many organizations start with the insurance value. Although this figure is usually not the exact answer, it is a good starting point. Some organizations have this information available in their financial group.

A derivative of this indicator is the stores investment (in spare parts) as a percentage of the estimated replacement value. This benchmark divides the total dollar valuation (in current dollars) for the maintenance spare parts by the estimated replacement value of the plant or facility. This value is almost always equivalent to half of the total maintenance costs divided by the estimated replacement value. This benchmark adds credibility to the maintenance budget guidelines discussed earlier in the textbook.

Another derivative of the estimated replacement value indicator is the value of assets maintained per maintenance technician. This benchmark is essentially the average dollar value of assets that each maintenance technician is responsible for maintaining. It is a fair benchmark because it does not involve ratios such as maintenance headcount to plant headcount. These benchmarks are summarized in Figure 12-1.

FIGURE 12-1 ASSET VALUE BASED BENCHMARKS

Indicator	Low Range	High Range	Best Practice
Maintenance Cost / ERV	2%	5%	2%
Stores Investment / ERV	.8%	1.2%	1%
ERV / Maintenance Technician	$4 M	$10 M	$7 M

BEST PRACTICE BENCHMARKS: MAINTENANCE STAFFING

Three staffing benchmarks are important to consider. The first is the maintenance technician to supervisor ratio. This ratio may range from 8 to 15 technicians per supervisor, with a Best Practice average being about 10 technicians per supervisor.

A second staffing ratio is the number of technicians to maintenance planner ratio. The range for this ratio should be 15 to 25 maintenance technicians per planner. The Best Practice average is about 20 maintenance technicians per planner. Anything above the 25:1 ratio could be disastrous for the planning program. In this case, the planner would no longer plan, but become an expediter. This level will not enhance the efficiency of the maintenance organization.

A third staffing benchmark looks at the estimated replacement value of the assets for which each maintenance engineer is responsible. In Figure 12-2, the ratio is quite broad, with a range of $50 million to $250 million per maintenance engineer. The Best Practice average is about $100 million per engineer. The range is broad because there is no consistent job description for a maintenance engineer. As this position becomes more clearly defined (See Chapter 3) in organizations, the ratio will approach the Best Practice average of $100 million per engineer. These numbers are summarized in Figure 12-2.

FIGURE 12-2 STAFFING BENCHMARKS

Indicator	Low Range	High Range	Best Practice
Technicians to Supervisor	8:1	15:1	10:1
Technicians to Planner	15:1	25:1	20:1
ERV / Maintenance Technician	$50 M	$250 M	$100 M

Maintenance and Sales Costs

Another way of examining maintenance cost is to compare it to the total cost of sales. This cost comparison is not as effective as comparing maintenance to the estimated replacement value of the asset. However, because many organizations do utilize this comparison, it is mentioned in this context. It is not a fair evaluation of the maintenance costs because the total cost of sales is variable and maintenance cannot always control the other variables in the cost calculation.

The range for this measure is from 1 to 5%, with a Best Practice average of about 2%. Similarly, the maintenance labor costs and the maintenance stores costs compared to sales costs are about 50% each. These figures are summarized in Figure 12-3

FIGURE 12-3 MAINTENANCE COSTS COMPARED TO TOTAL COST OF SALES

Indicator	Low Range	High Range	Best Practice
Total Maintenance Costs/Sales Cost	8	5%	2%
Maintenance Labor Costs / Sales Costs	.6%	2.5%	1%
Maintenance Stores Costs / Sales Costs	.4%	2.5%	1%

MAINTENANCE PERFORMANCE

The next series of benchmarks are related to maintenance performance. The first, work order coverage, represents the percent of maintenance work that actually is reported to a maintenance work order. Benchmarks range from 60 to 100%. The Best Practice would actually record 100% of all work performed for future reference and reporting analysis.

The next benchmark is the preventive maintenance compliance benchmark. Its range is from 65 to 100% completion rate. The Best Practice would be to complete 100% of all the preventive maintenance tasks that are scheduled. If anything less is accomplished, the impact would be seen on equipment availability.

For the maintenance schedule compliance benchmark, the indicator range goes 35 to 95%. The Best Practice benchmark is 95%. Although it may seem that 100% should be the benchmark, an organization is unlikely to ever achieve 100%, due to the reactive work that is encountered during the execution of the weekly schedule. An organization that achieves 95% has achieved a Best Practice. These benchmarks are summarized in Figure 12-4.

FIGURE 12-4 MAINTENANCE PERFORMANCE

Indicator	Low Range	High Range	Best Practice
Work Order Coverage	60%	100%	100%
Preventive Maintenance Compliance	65%	100%	100%
Maintenance Schedule Compliance	35%	95%	95%

Another maintenance performance benchmark is planned maintenance work, maintenance activities that are planned, rather than reactive. The benchmarks range from 35 to 95%. The Best Practice range is above 80%. This range is not more specific because the type of business dictates the amount of planned work that can be accomplished. As long as less than 20% of the work is reactive, which means 80% or more is planned and scheduled, the organization is considered to be in a Best Practice category.

Another maintenance performance benchmark is the amount of operator involvement in the preventive maintenance program. The benchmarks range from 10 to 40%. The Best Practice benchmark actually varies because the type of work, the type of equipment, the operator skill level, safety and health regulations, and the union agreements vary. All of these variables impact the amount of involvement that operators can have in preventive maintenance activities. The goal of operator involvement is to free up some maintenance resources for redeployment in the predictive and condition based activities. As long as this is achieved to the optimum, an organization can be considered in the Best Practice category.

Another area that is examined under maintenance performance is the amount of contract maintenance that is performed. Benchmarks show a range of contractor usage from 10 to 100%. The Best Practice percentage varies depending on the business needs. In some organizations, a business decision is made to contract out more of the maintenance. In other organizations, it is preferred to keep the maintenance activities in-house. Whatever decision is made, it should be based on financial parameters. As long as the decision is properly made, the organization has achieved the optimum balance of contract versus in-house maintenance and is at its Best Practice. These benchmarks are summarized in Figure 12-5.

A third group of indicators that examine maintenance performance is found in Figure 12-6. The first of these indicators is the percentage of hours spent on preventive and predictive maintenance activities compared to the total hours worked. The benchmarks range from 20 to 50%. The Best Practice average is 50%, which allows for 30 to 40% additional corrective work and less than 20% reactive work.

FIGURE 12-5 MAINTENANCE PERFORMANCE: 2

Indicator	Low Range	High Range	Best Practice
Planned Maintenance Work	35%	95%	80+%
Operator Involvement In PM	10%	40%	Varies
Contractor Costs / Total Maintenance Costs	10%	100%	Varies

A second indicator in this group is the percent of reactive hours compared to total hours. This indicator ranges from 5 to 50% or more. A Best Practice is less than 10% reactive work. Most organizations are doing quite well if they have less than 20% reactive work. However, they should not be satisfied with these results. Less than 10% is the optimum number. If an organization has achieved this low percentage of reactive work, it is considered to be a Best Practice organization.

The last indicator under maintenance performance is the productivity rate for what is commonly referred to as a wrench time. This is the amount of hands-on time a technician spends working per hour. The benchmarks show a low range of 20% for reactive organizations to a high range of 60% for proactive organizations. The more planned and scheduled an organization is, the closer it will come to the Best Practice percentage of 60%. An organization with 20% wrench time is spending about three times the amount of money that it should to accomplish the same work that a Best Practice organization could accomplish. These statistics are summarized in Figure 12-6.

FIGURE 12-6 MAINTENANCE PERFORMANCE: 3

Indicator	Low Range	High Range	Best Practice
PM/PDM Hours/ Total Hours	20%	50%	50%
Reactive Hours/ Total Hours	5%	50%	<10%
Productivity Rates (Wrench Time)	20%	60%	60%

EQUIPMENT PERFORMANCE

The next series of indicators, which focus on equipment performance, are categorized under maintenance because maintenance impacts the performance of the asset. The first of these indicators is equipment availability. This benchmark ranges from 65 to 99.9%. In reality, the Best Practice percentage varies because the higher the equipment availability is, the more expensive it is. Therefore, the cost of production should be used to determine the level of availability required. If a company strives for 99.9% availability when only 90% availability is needed, then it is spending too much for its maintenance program. The Best Practice percentage must be determined by each company for each piece of equipment or process.

The next benchmark is equipment efficiency. It compares the actual output of a piece of equipment to its design output. This benchmark ranges from 75 to 95%. The Best Practice percentage is 95% or more. As with equipment performance, whether or not 95% is sufficient depends on the product. It may not be cost effective to try to

FIGURE 12-7 EQUIPMENT PERFORMANCE			
Indicator	**Low Range**	**High Range**	**Best Practice**
Equipment Availability	65%	99.9%	**Varies**
Equipment Efficiency	75%	95%	95+%
Overall Equipment Effectiveness	<20%	85+%	**Varies**

gain the last percent or two of efficiency from a piece of equipment. That decision is one that company officials must make. But 95% is a Best Practice threshold.

The final benchmark in this series is the overall equipment effectiveness. The low range on overall equipment effectiveness has been observed at less than 20% with a high range of more than 85%. The Best Practice percentage varies depending upon the organization's needs. For some processes, companies spend too much to get to 85% and the decision is not cost justifiable. In other organizations, 85% is too low of a percentage. All business factors must be considered when setting the Best Practice threshold for a particular process or piece of equipment. These figures are summarized in Figure 12-7.

MAINTENANCE STORES

The next series of benchmarks focuses on maintenance stores. The first benchmark, the maintenance spare parts to inventory turns ratio, ranges from 0.5 to 1.4. The Best Practice benchmark varies, but trends closer to 1.4. Factors that determine this range include foreign equipment, availability of spare parts, and location of suppliers. Any of these factors can drive the turns ratio to the low range.

A second related benchmark is the stores service level which ranges from 80 to 99%. A Best Practice value is between 95 and 97%. This value is based on a business decision. A service level below 95% will result in unnecessary downtime due to parts outages. A service level above 97% usually indicates that too many spare parts are being carried. The proper balance for determining the Best Practice level depends on financial considerations of downtime versus holding cost.

The last benchmark in this category is the value of stores transactions per stores personnel. This benchmark indicates whether or not a storeroom is properly staffed. The benchmark ranges from $350,000 to $600,000 per year. The Best Practice benchmark varies based on a business decision. This decision involves factors such as storeroom locations, production shifts serviced, maintenance shifts serviced, and location. In some organizations, where the maintenance stores organization does not match the maintenance organization, the lower range

FIGURE 12-8 MAINTENANCE STORES

Indicator	Low Range	High Range	Best Practice
Spare Parts Inventory Turns	.5	1.4	Varies
Stores Service Level	80%	99%	95–97%
Value of Stores Transactions per Store Personnel	$350K	$600K+	Varies

is an acceptable number. In a location that is optimized, the storeroom attendant would be able to issue a higher dollar value of spare parts. Because there is no definitive benchmark in this category, these figures indicate possible ranges. These benchmarks are presented in Figure 12-8.

Maintenance Training

The final group of benchmarks look at maintenance training. The first benchmark is training expenditures per employee. This expenditure ranges from $607 to $2000 per employee. The Best Practice varies on this indicator as it will on the following ones. This expenditure is based on the current skills of the employees and their identified training needs. If an organization has a highly trained workforce and is not currently modernizing the plant at the current time, the low range of training dollars may be acceptable. However, if an organization has identified skills deficiencies and is modernizing its plant, the high range may be insufficient. This benchmark needs to be balanced against both the current business climate and the condition of the employee skills at the plant.

The second benchmark is training as a percentage of payroll; it ranges from 1.65 to 4.39%. As noted above, the Best Practice percentage will vary.

The last benchmark measures technical training as a percent of total training expense. It ranges from less than 20 to more than 50%. The Best Practice percentage will vary. Technical training should not be sacrificed for compliance or soft skills training. It should be specified, based on the skills deficiencies noted as well as new equipment being brought into the plant. These benchmarks are summarized on Figure 12-9.

FIGURE 12-9 TRAINING

Indicator	Low Range	High Range	Best Practice
Training Expenditure per Employee	$607	$2000	Varies
Training as a Percentage of Payroll	1.65%	4.39%	Varies
Technology Training / Total Training Expense	<20%	50+%	Varies

Conclusion

Why should we benchmark? The gap between present practices and Best Practices promotes dissatisfaction and the desire for change. Visiting a benchmarking partner — seeing, understanding, and learning from the Best Practice — helps to identify what you can change and how to change; it provides a realistic, yet achievable picture of what the future could be. Without this vision, many people will never fully comprehend the direction that would mean for the success of the company.

Organizations considering change always face the argument "We have always done it is way." Yet why should this argument be a reason for continuing to do something the same way? Annual objectives based solely on past performance plus or minus 10% are meaningless and destructive in a period of rapid change.

It is necessary to explore the tangible and intangible factors that combine to produce a superior performer. Those people who are most directly concerned in the activity being benchmarked should be involved. Without this level of involvement, these individuals are unlikely to embrace the vision. Without their support, no benchmarking exercise will ever be truly successful.

Benchmarking is not without its own limitations. Benchmarks can be too fluid because world standards are rapidly improving. They are often too modest for corporate goals. Too often benchmarks simply become numbers to achieve, and the real goal — the opportunity for continuous and rapid improvement — quickly diminishes.

In order to gain the maximum benefit from benchmarking, four factors must be remembered:

- A benchmark provides a measure for the benchmark process among the benchmarking partners.
- A benchmark describes the organization's gap in performance compared to the measure of the benchmark partners.
- A benchmark identifies the Best Practices and enablers that produced results that were observed during the benchmarking study.
- A benchmark sets performance goals for the benchmarked process and identifies actions that can be taken to improve performance.

A benchmark performance does not remain standard for long. In this age of continuous and rapid improvement, a benchmark today will be a standard tomorrow, and a mediocre performance in the future. Only by seeing benchmarks as part of a continuous improvement program will any company make the progress necessary to stay in business.

Benchmarks, therefore, are good for finding process improvements, arousing people to a challenge, and setting milestone targets. They are not the end-all of the business, nor should they ever be. If a company fails to learn during a benchmarking project, it likely made mistakes. It will not be successful in future benchmarking endeavors without reviewing the project and looking for, finding, and correcting the mistakes.

An additional mistake made when benchmarking is conducting an extensive analysis beforehand. This delays improvements. An extensive upfront audit is often a way people postpone doing something useful. Benchmarking does not require this audit. Many changes can be quickly identified and implemented. Benefits can be derived in a rapid fashion.

In conclusion, ask yourself the following questions:

- Will you help your company achieve the benefits available by improving maintenance?

- Will you utilize benchmarking as a tool for continuous improvement?

- Will you insure that you benchmark for the right reasons?

If your answer to these questions is yes, then you will be increasing your company's competitive position and insuring its future.

INDEX

www.ingramcontent.com/pod-product-compliance
Lightning Source LLC
Chambersburg PA
CBHW071252180926
51613CB00001B/16
* 9 7 8 0 8 3 1 1 3 5 0 3 4 *